By Ed McCarthy and Mary Ewing-Mulligan

1. **Wine 101** 4
The basics: turning grapes into wine and telling red from white

Feature 1: Keeping Leftover Wine 9

2. **Grape Expectations: Grape Varieties** 10
Sorting out the raw materials: grapes and what they do

3. **Judging a Wine by Its Label** 16
Decoding the label lingo

4. **The Wine Name Game** 22
Understanding what's in a name through the mysteries of terroir and brand

5. **These Taste Buds Are for You** 28
Getting to grips with the subject: smelling, tasting and describing wine

6. **Finding and Buying Good Wine** 34
Knowing your good wine from your bad, and where to find the good stuff

7. **Marrying Wine with Food** 40
Matching what you drink with what you want to eat, whether dining in or out

Feature 2: Restaurant Wine Tips 46

8. **French Wines and the Legendary Bordeaux** ... 48
Taking a trip through the heartland of great wine

9. **Burgundy: The Other Great French Wine** 56
From Beaujolais to Chablis – and beyond

10. **Other French Wines** 62
Discovering the riches of the Rhône and the Loire – amongst others

11. **Wine in Italy: Piedmont and Tuscany** 68
Getting in the zone with Italy's great reds

12. **Tre Venezie and Other Italian Regions** 74
Exploring the northeast - and the rest

Feature 3: Does the Glass Really Matter? 78

13. **Intriguing Wines from Old Spain** 80
Running the rule over Rioja, Ribera del Duero and Priorato

14. **Wines in Germany and Portugal** 86
Reviewing Germany's Rieslings and Portugal's Vinho Verde

15. **Wines Elsewhere in Europe** 92
Rounding up the best of the rest of Europe

16. **North American Wines** 96
Checking our the best in the West

Feature 4: Does Wine Really Breathe? 102

17. **Australian and New Zealand Wines Arise** ... 104
Discovering delights down under, from Murray River to Marlborough

18. **Heating It Up with Chile, Argentina, and South Africa** 110
Surveying the Southern hemisphere in Old World and the New

19. **Champagnes and Other Sparkling Wines** 116
Adding some sparkle with Champagne and other bubblies

20. **Wine Roads Less Travelled: Fortified and Dessert Wines** 122
Coming on strong with fortified and dessert wines

Feature 5: Vintage Wine Chart 128

Exploring Wine For Dummies®
Published by
John Wiley & Sons, Ltd
The Atrium
Southern Gate
Chichester
West Sussex
PO19 8SQ
England

E-mail (for orders and customer service enquires): **cs-books@wiley.co.uk**

Visit our Home Page on **www.wiley.com**

For general information on our other products and services, please contact our Customer Care Department within the U.S. at 800-762-2974, outside the U.S. at 317-572-3993, or fax 317-572-4002.

For technical support, please visit **www.wiley.com/techsupport**.

Wiley also publishes its books in a variety of electronic formats. Some content that appears in print may not be available in electronic books.
British Library Cataloguing in Publication Data: A catalogue record for this book is available from the British Library

ISBN: 978-0-470-97875-7

Printed and bound in Great Britain by Stones the Printers, Banbury

10 9 8 7 6 5 4 3 2 1

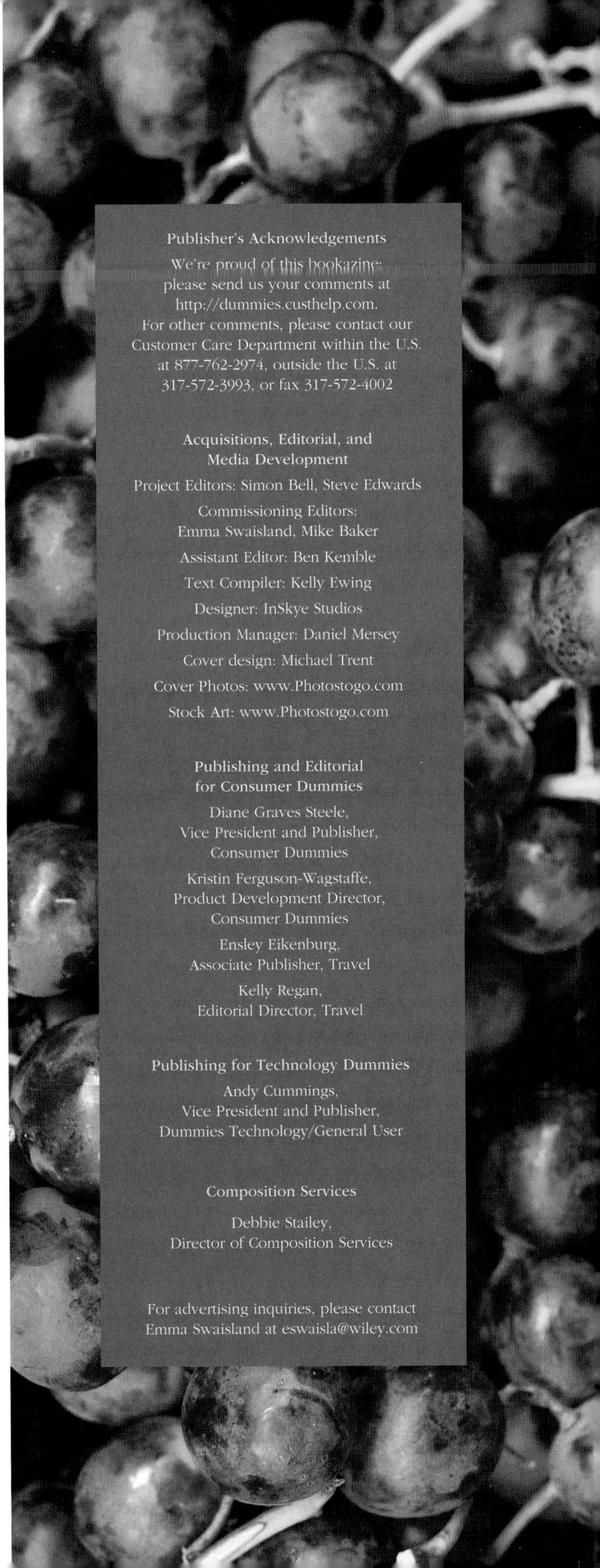

Publisher's Acknowledgements

We're proud of this bookazine; please send us your comments at http://dummies.custhelp.com.
For other comments, please contact our Customer Care Department within the U.S. at 877-762-2974, outside the U.S. at 317-572-3993, or fax 317-572-4002

Acquisitions, Editorial, and Media Development

Project Editors: Simon Bell, Steve Edwards

Commissioning Editors: Emma Swaisland, Mike Baker

Assistant Editor: Ben Kemble

Text Compiler: Kelly Ewing

Designer: InSkye Studios

Production Manager: Daniel Mersey

Cover design: Michael Trent

Cover Photos: www.Photostogo.com

Stock Art: www.Photostogo.com

Publishing and Editorial for Consumer Dummies

Diane Graves Steele, Vice President and Publisher, Consumer Dummies

Kristin Ferguson-Wagstaffe, Product Development Director, Consumer Dummies

Ensley Eikenburg, Associate Publisher, Travel

Kelly Regan, Editorial Director, Travel

Publishing for Technology Dummies

Andy Cummings, Vice President and Publisher, Dummies Technology/General User

Composition Services

Debbie Stailey, Director of Composition Services

For advertising inquiries, please contact Emma Swaisland at eswaisla@wiley.com

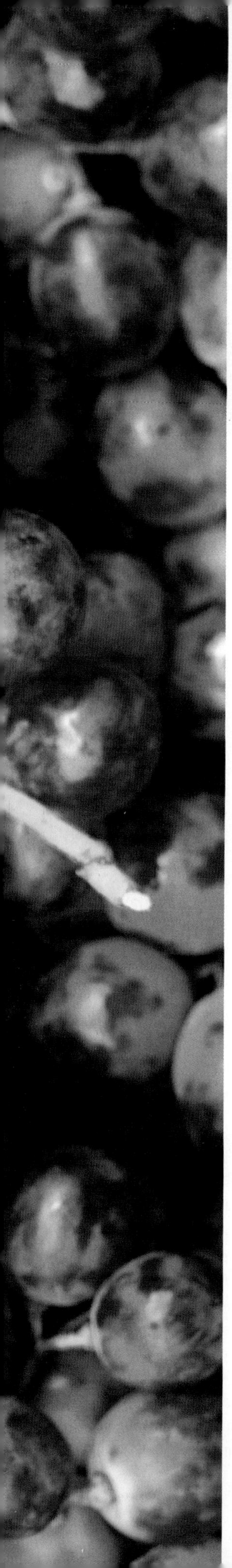

Introduction

We love wine. We love the way it tastes, we love the fascinating variety of wines in the world, and we love the way wine brings people together at the dinner table. We believe that you and everyone else should be able to enjoy wine – regardless of your experience or your budget.

But we'll be the first to admit that wine people, such as many wine professionals and really serious connoisseurs, don't make it easy for regular people to enjoy wine. You have to know strange names of grape varieties and foreign wine regions. You have to figure out whether to buy a £20-a-bottle wine or a £5-a-bottle wine that seem to be pretty much the same thing. You even need a special tool to open the bottle once you get it home!

All this complication surrounding wine will never go away, because wine is a very rich and complex field. But you don't have to let the complication stand in your way. With the right attitude and a little understanding of what wine is, you can begin to buy and enjoy wine. And if, like us, you decide that wine is fascinating, you can find out more and turn it into a wonderful hobby.

Because we hate to think that wine, which has brought so much pleasure into our lives, could be the source of anxiety for anyone, we want to help you feel more comfortable around wine. Some knowledge of wine, gleaned from the pages of this book and from our shared experiences, will go a long way toward increasing your comfort level.

Ironically, what will *really* make you feel comfortable about wine is accepting the fact that you'll never know it all – and that you've got plenty of company.

You see, after you really get a handle on wine, you discover that *no one* knows everything there is to know about wine. There's just too much information, and it's always changing. And when you know that, you can just relax and enjoy the stuff.

Wine 101

IN THIS ARTICLE

- *Transforming grapes into wine*
- *Finding out what red wine has that white wine doesn't*
- *Discovering the differences between table wine, fortified wine, and sparkling wine*

Plenty of people enjoy drinking wine but don't know much about it. Knowing a lot of information about wine definitely isn't a prerequisite to enjoying it. But familiarity with certain aspects of wine can make choosing wines a lot easier, enhance your enjoyment of wine, and increase your comfort level.

How Wine Happens

Wine is, essentially, nothing but liquid, fermented fruit. Most wines in the world – 99.9 per cent – are made from grapes, but you can substitute raspberries or any other fruit.

After the grapes are crushed, *yeasts* (tiny one-celled organisms that exist naturally in the vineyard and, therefore, on the grapes) come into contact with the sugar in the grapes' juice and gradually convert that sugar into alcohol. Yeasts also produce carbon dioxide, which evaporates into the air. When the yeasts are done working, your grape juice is wine. The sugar that was in the juice is no longer there – alcohol is present instead. (The riper and sweeter the grapes, the more alcohol the wine will have.) This process is called *fermentation*.

If every winemaker actually made wine in this crude a manner, you'd be drinking some pretty rough stuff that would hardly inspire someone to write a wine book. But today's winemakers have a bag of tricks as big as a sumo wrestler's appetite. That's one reason why no two wines ever taste exactly the same.

Winemakers can control the type of container they use for the fermentation process (stainless steel and oak are the two main materials), as well as the size of the container and the temperature of the juice during fermentation – and every one of these choices can make a big difference in the taste of the wine. After fermentation, they can choose how long to let the wine *mature* (a stage when the wine sort of gets its act together) and in what kind of container. Fermentation can last three days or three months, and the wine can then mature for a couple of weeks or a couple of years or anything in between.

TIP

If you have trouble making decisions, don't ever become a winemaker.

Local flavour

Obviously, one of the biggest factors in making one wine different from the next is the nature of the raw material, the grape juice. Different varieties of grapes (Chardonnay, Cabernet Sauvignon, or Merlot, for example) make different wines.

Grapes, the raw material of wine, don't grow in a void. Where they grow — the soil and climate of each wine region, as well as the traditions and goals of the people who grow the grapes and make the wine — affects the nature of the ripe grapes, and the taste of the wine made from those grapes.

What Colour Is Your Appetite?

Your inner child will be happy to know that when it comes to wine, it's okay to like some colours more than others. You can't get away with saying "I don't like green food!" much beyond your sixth birthday, but you can express a general preference for white, red, or pink wine for all your adult years.

(Not exactly) white wine

Whoever coined the term "white wine" must have been colourblind. All you have to do is look at it to see that it's not white, it's yellow.

White wine is wine without any red colour (or pink colour, which is in the red family). ***White Zinfandel,*** a popular pink wine, isn't white wine. But yellow wines, golden wines, and wines that are as pale as water are all white wines.

Wine becomes white wine in one of two ways:

- White wine can be made from white grapes – which, by the way, aren't white. *White* grapes are greenish, greenish yellow, golden yellow, or sometimes even pinkish yellow. Basically, white grapes include all the grape types that are not dark red or dark bluish. If you make a wine from white grapes, it's a white wine.

- The second way a wine can become white is a little more complicated. The process involves using red grapes – but only the *juice* of red grapes, not the grape skins. The juice of most red grapes has no red pigmentation – only the skins do – and so a wine made with only the juice of red grapes can be a white wine. In practice, though, very few white wines come from red grapes. (Champagne is one exception.)

TIP

Serve white wines cool, but not ice-cold. Sometimes restaurants serve white wines too cold, and you have to wait for the wine to warm up before you drink it. If you usually drink your wine cold, try drinking your favourite white wine a little less cold sometime, and you'll discover it has more flavor that way.

Is white always right?

You can drink white wine anytime you like – which, for most people, means as a drink without food or with lighter foods.

White wines are often considered *apéritif* wines, meaning wines consumed before dinner, in place of cocktails, or at parties. (If you ask the officials who busy themselves defining such things, an apéritif wine is a wine that has flavours added to it, as vermouth does. But unless you're in the business of writing wine labels for a living, don't worry about that.)

A lot of people like to drink white wines when the weather is hot because they're more refreshing than red wines, and they're usually drunk chilled (the wines, not the people).

There's no such thing as plain white wine

White wines fall into four general taste categories, not counting sparkling wine or the really sweet white wine that you drink with dessert. Here are the four broad categories:

- Some white wines are *fresh, unoaked whites* – crisp and light, with no sweetness and no oaky character. Most Italian white wines, like Soave and Pinot Grigio, and some French whites, like Sancerre and some Chablis wines, fall into this category.
- Some white wines are *earthy whites* – dry, fuller-bodied, unoaked or lightly oaked, with a lot of earthy character. Some French wines, such as Mâcon or whites from the Côtes du Rhône region, have this taste profile.
- Some white wines are *aromatic whites* – characterized by intense aromas and flavours that come from their particular grape variety, whether they're *off-dry* (that is, not bone-dry) or dry. Examples include a lot of German wines, and wines from flavourful grape varieties such as Riesling or Viognier.
- Finally, some white wines are rich, oaky whites – dry or fairly dry, and full-bodied with pronounced oaky character. Most Chardonnays and many French wines – like many of those from the Burgundy region of France – fall into this group.

Red, red wine

In this case, the name is correct: Red wines really are red. They can be purple red, ruby red, or garnet, but they're red.

Red wines are made from grapes that are red or bluish in colour. So guess what wine people call these grapes? Black grapes!

The most obvious difference between red wine and white wine is colour. The red colour occurs when the colourless juice of red grapes stays in contact with the dark grape skins during fermentation and absorbs the skins' colour. Along with colour, the grape skins give the wine *tannin,* a substance that's an important part of the way a red wine tastes. The presence of tannin in red wines is actually the most important taste difference between red wines and white wines.

Red wines vary quite a lot in style, partly because winemakers have so many ways of adjusting their red-winemaking to achieve the kind of wine they want. For example, if winemakers leave the juice in contact with the skins for a long time, the wine becomes more *tannic* (firmer in the mouth, like strong tea; tannic wines can make you pucker). If winemakers drain the juice off the skins sooner, the wine is softer and less tannic.

Thanks to the wide range of red wine styles, you can find red wines to go with just about every type of food and every occasion when you want to drink wine (except the times when you want to drink a wine with bubbles, because most bubbly wines are white or pink).

TIP

Red wine tends to be consumed more often as part of a meal than as a drink on its own.

There's no such thing as just plain red wine. Here are four red wine styles:

- *Soft, fruity reds* are relatively light-bodied, with a lot of fruitiness and little tannin (like Beaujolais Nouveau wine from France, some Valpolicellas from Italy, and many cheaper U.S. wines).
- *Mild-mannered reds* are medium-bodied with subtle, un-fruity flavours (like less expensive wines from Bordeaux, in France and some inexpensive Italian reds).
- *Spicy reds* are flavourful, fruity wines with spicy accents and some tannin (such as some Malbecs from France or Argentina, and Dolcettos from Italy).
- *Powerful reds* are full-bodied and tannic (such as the most expensive California Cabernets; Barolo, from Italy; the most expensive Australian reds; and lots of other expensive reds).

WARNING!

One sure way to spoil the fun in drinking most red wines is to drink them too cold. Those tannins can taste really bitter when the wine is cold – just as in a cold glass of very strong tea. On the other hand, many restaurants serve red wines too warm. If the bottle feels cool to your hand, that's a good temperature.

Are there any wines without sulphites?

Sulphur dioxide exists naturally in wine as a result of fermentation. It also exists naturally in other fermented foods, such as bread, cookies, and beer. (Various sulphur derivatives are also used regularly as preservatives in packaged foods.)

Winemakers use sulphur dioxide at various stages of the winemaking process because it stabilizes the wine (preventing it from turning to vinegar or deteriorating from oxygen exposure) and safeguards its flavour. Sulphur has been an important winemaking tool since Roman times.

Very few winemakers refrain from using sulphur dioxide, but some do. Your wine shop may carry a few wines whose sulfite content is so low that their labels do not have to carry the phrase *Contains Sulphites*.

Some people complain that they can't drink red wines without getting a headache or feeling ill. Usually, they blame the sulphites in the wine. Red wines, however, contain far less sulphur than white wines. That's because the tannin in red wines acts as a preservative, making sulphur dioxide less necessary. Red wines do contain histamine-like compounds and other substances derived from the grape skins that could be the culprits. Whatever the source of the discomfort, it's probably not sulphites.

So if you wish to limit your consumption of sulphites, dry red wines should be your first choice, followed by dry white wines. Sweet wines contain the most sulphur dioxide.

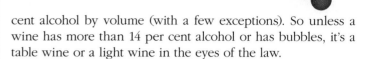

A rose is a rose, but a rosé is "white"

Rosé wines are pink wines. Rosé wines are made from red grapes, but they don't end up red because the grape juice stays in contact with the red skins for a very short time – only a few hours, compared to days or weeks for red wines. Because this *skin contact* (the period when the juice and the skins intermingle) is brief, rosé wines absorb very little tannin from the skins. Therefore, you can chill rosé wines and drink them as you would white wines.

Of course, not all rosé wines are called rosés. Many rosé wines today are called *blush* wines – a term invented by wine marketers to avoid the word *rosé,* because back in the '80s, pink wines weren't very popular. Lest someone figures out that *blush* is a synonym for *rosé,* the labels call these wines *white.* But even a child can see that White Zinfandel is really pink.

The blush wines that call themselves *white* are fairly sweet. Wines labeled *rosé* can be sweetish, too, but some wonderful rosés from Europe (and a few from America, too) are *dry* (not sweet). Some hardcore wine lovers hardly ever drink rosé wine, but many wine drinkers are discovering what a pleasure a good rosé wine can be, especially in warm weather.

Other Ways of Categorizing Wine

Regular wines – red, white, or pink – without bubbles are called *light* wines in Europe and *table* wines in America. Sometimes they're even referred to as *still* wines because they don't have bubbles moving around in them.

Table wine

Table wine, or light wine, is fermented grape juice whose alcohol content falls within a certain range. Furthermore, table wine is not bubbly. (Some table wines have a very slight carbonation, but not enough to disqualify them as table wines.) According to U.S. standards of identification, table wines may have an alcohol content no higher than 14 per cent; in Europe, light wine must contain from 8.5 per cent to 14 per cent alcohol by volume (with a few exceptions). So unless a wine has more than 14 per cent alcohol or has bubbles, it's a table wine or a light wine in the eyes of the law.

TIP

Here's a real-world definition of table wines: They're the normal, nonbubbly wines that most people drink most of the time.

Liqueur wine

Many wines have more than 14 per cent alcohol because the winemaker added alcohol during or after the fermentation. That's an unusual way of making wine, but some parts of the world, like the Sherry region in Spain and the Port region in Portugal, have made quite a specialty of it.

In Europe, this category of wines is called *liqueur wines,* which carries the same connotation of sweetness. Another term is *fortified,* which suggests that the wine has been strengthened with additional alcohol.

WARNING!

Dessert wine is the legal terminology for these sweet wines, probably because they're usually sweet and often enjoyed after dinner. That term is misleading because dessert wines aren't *always* sweet and aren't *always* consumed after dinner. (Dry Sherry is categorized as a dessert wine, for example, but it's dry, and some people drink it before dinner.)

Sparkling wine

Sparkling wines are wines that contain carbon dioxide bubbles. Carbon dioxide gas is a natural byproduct of fermentation, and winemakers sometimes decide to trap it in the wine. Just about every country that makes wine also makes sparkling wine.

In Europe, the United States, and Canada, *sparkling wine* is the official name for the category of wines with bubbles. Isn't it nice when everyone agrees?

Champagne (with a capital C) is the most famous sparkling wine – and probably the most famous *wine,* for that matter. Champagne is a specific type of sparkling wine (made from certain grape varieties and produced in a certain way) that comes from a region in France called Champagne. It's the undisputed Grand Champion of Bubblies.

Unfortunately for the people of Champagne their wine is so famous that the name *champagne* has been borrowed again and again by producers elsewhere, until the word has become synonymous with practically the whole category of sparkling wines. For example, until a recent agreement between the United States and the European Union, U.S. winemakers could legally call any sparkling wine *champagne* – even with a capital *C,* if they wanted – as long as the carbonation was not added artificially. Even now, those American wineries that were already using that name may continue to do so. (They do have to add a qualifying geographic term such as *American* or *Californian* before the word Champagne, however.)

For the French, limiting the use of the name *champagne* to the wines of the Champagne region is a *cause célèbre.* European Union regulations not only prevent any other member country from calling its sparkling wines *champagne* but also prohibit the use of terms that even *suggest* the word *champagne,* such as fine print on the label saying that a wine was made by using the "champagne method." What's more, bottles of sparkling wine from countries outside the European Union that use the word champagne on the label are banned from sale in Europe. The French are that serious.

Keeping Leftover Wine

A sparkling-wine stopper, a device that fits over an opened bottle, is really effective in keeping any remaining Champagne or sparkling wine fresh (often for several days) in the refrigerator. But what do you do when you have red or white wine left in the bottle?

You can put the cork back in the bottle if it still fits, and put the bottle into the refrigerator. (Even red wines will stay fresher there; just take the bottle out to warm up about an hour before serving it.) But four other methods are also reliable in keeping your remaining wine from oxidizing. These techniques are all the more effective if you put the bottle in the fridge after using them:

- If you have about half a bottle of wine left, simply pour the wine into a clean, empty half-sized wine bottle and recork the smaller bottle. We sometimes buy wines in half-bottles, just to make sure that we have the empty half-bottles around.

- Use a handy, inexpensive, miniature pump called a Vacuvin in most wine stores. This pump removes oxygen from the bottle, and the rubber stoppers that come with it prevent additional oxygen from entering the bottle. It's supposed to keep your wine fresh for up to a week, but it doesn't always work that well, in our experience.
- Buy small cans of inert gas in some wine shops . Just squirt a few shots of the gas into the bottle through a skinny straw, which comes with the can, and put the cork back in the bottle. The gas displaces the oxygen in the bottle, thus protecting the wine from oxidizing. Simple and effective. Private Preserve is a good brand, and highly recommended.
- A new device, called WineSavor, is a flexible plastic disk that you roll up and insert down the bottle's neck. Once inside the bottle, the disk opens up and floats on top of the wine, blocking the wine from oxygen.

To avoid all this bother, just drink the wine! Or, if you're not too fussy, just place the leftover wine in the refrigerator and drink it in the next day or two – before it goes into a coma.

Grape Expectations: Grape Varieties

IN THIS ARTICLE

- *Sorting out the traits of grapes*
- *Exploring white grape varieties*
- *Talking about red grape varieties*

Grapes are the starting point of every wine, and therefore they're largely responsible for the style and personality of each wine. Grapes also are one of the easiest ways to classify wine and make sense of the hundreds of different types of wine that exist.

REMEMBER

The specific grape variety (or varieties) that makes any given wine is largely responsible for the sensory characteristics the wine offers — from its appearance to its aromas, its flavours, and its alcohol–tannin–acid profile. How the grapes grow — the amount of sunshine and moisture they get, for example, and how ripe they are when they're harvested — can emphasize certain of characteristics rather than others. So can winemaking processes such as oak aging. Each grape variety reacts in its own way to the farming and winemaking techniques that it faces.

REMEMBER

Skin colour is the most fundamental distinction among grape varieties. Every grape variety is considered either a white variety or a red (or "black") one, according to the colour of its skins when the grapes are ripe. (A few red-skinned varieties are further distinguished by having red pulp rather than white pulp.)

What a Personality! Personality Traits of Grape Varieties

Grape variety refers to the fruit of a specific type of grapevine – the fruit of the Cabernet Sauvignon vine, for example, or of the Chardonnay vine. *Personality traits* are the characteristics of the fruit itself – its flavours, for example.

Individual grape varieties also differ from one another in other ways:

- **Aromatic compounds:** Some grapes (like Muscat) contribute floral aromas and flavours to their wine, for example, while other grapes contribute herbaceous notes (as Sauvignon Blanc does) or fruity character. Some grapes have very neutral aromas and flavours and, therefore, make fairly neutral wines.
- **Acidity levels:** Some grapes are naturally disposed to higher acid levels than others, which influences the wine made from those grapes.
- **Thickness of skin and size of the individual grapes (called *berries*):** Black grapes with thick skins naturally have more tannin than grapes with thin skins; ditto for small-berried varieties compared to large-berried varieties, because their skin-to-juice ratio is higher. More tannin in the grapes translates into a firmer, more tannic red wine.

The composite personality traits of any grape variety are fairly evident in wines made from that grape. A Cabernet Sauvignon wine is almost always more tannic and slightly lower in alcohol than a comparable Merlot wine, for example, because that's the nature of those two grapes.

Built for Speed: The Performance Factors of Grape Varieties

Performance factors refer to how the grapevine grows, how its fruit ripens, and how quickly it can get from 0 to 60 miles per hour.

The performance factors that distinguish grape varieties are vitally important to the grape grower because those factors determine how easy or challenging it will be for him to cultivate a specific variety in his vineyard – if he can even grow it at all. The issues include

- How much time a variety typically needs to ripen its grapes
- How dense and compact the bunches of grapes are
- How much vegetation a particular variety tends to grow

A Primer on White Grape Varieties

This section includes descriptions of the most important white *vinifera* varieties today, as well as the types of wine that are made from each grape. These wines can be varietal wines or place-name wines that don't mention the grape variety anywhere on the label (a common practice for European wines). These grapes can also be blending partners for other grapes, in wines made from multiple grape varieties.

Chardonnay

Chardonnay is a regal grape for its role in producing the greatest dry white wines in the world — white Burgundies — and for being one of the main grapes of Champagne. Today, Chardonnay also ends up in a huge amount of everyday wine.

The Chardonnay grape grows in practically every wine-producing country of the world, for two reasons:

- It's relatively adaptable to a wide range of climates.
- The name Chardonnay on a wine label is, these days, a surefire sales tool.

Because the flavours of Chardonnay are very compatible with those of oak, most Chardonnay wine receives some oak treatment either during or after fermentation. Except for Northeastern Italy and France's Chablis and Mâconnais districts, where oak is usually not used for Chardonnay, oaky Chardonnay wine is the norm, and unoaked Chardonnay is the exception.

Chardonnay itself has fruity aromas and flavours that range from apple – in cooler wine regions – to tropical fruits, especially pineapple, in warmer regions. Chardonnay also can display subtle earthy aromas, such as mushroom or minerals. Chardonnay wine has medium to high acidity and is generally full-bodied. Classically, Chardonnay wines are dry. But most inexpensive Chardonnays these days are actually a bit sweet.

Chardonnay is a grape that can stand on its own in a wine, and the top Chardonnay-based wines (except for Champagne and similar bubblies) are 100 per cent Chardonnay.

TIP

Oaked Chardonnay is so common that some wine drinkers confuse the flavour of oak with the flavour of Chardonnay. If your glass of Chardonnay smells or tastes toasty, smoky, spicy, vanilla-like, or butterscotch-like, that's the oak you're perceiving, not the Chardonnay!

Riesling

The great Riesling wines of Germany have put the Riesling grape on the charts as an undisputedly noble variety. Riesling shows its real class only in a few places outside of Germany, however. The Alsace region of France, Austria, and the Clare Valley region of Australia are among the few.

Riesling wines are far less popular today than Chardonnay. Maybe that's because Riesling is the antithesis of Chardonnay. While Chardonnay is usually gussied up with oak, Riesling almost never is; while Chardonnay can be full-bodied and rich, Riesling is more often light-bodied, crisp, and refreshing. Riesling's fresh, vivid personality can make many Chardonnays taste clumsy in comparison.

High acidity, low to medium alcohol levels, and aromas/flavours that range from ebulliently fruity to flowery to minerally are trademarks of Riesling.

TIP

The common perception of Riesling wines is that they're sweet, and many of them are – but plenty of them aren't. Alsace Rieslings are normally dry, many German Rieslings are fairly dry, and a few American Rieslings are dry.

Sauvignon Blanc

Sauvignon Blanc is a white variety with a very distinctive character. It's high in acidity with pronounced aromas and flavours. Besides herbaceous character (sometimes referred to as *grassy*), Sauvignon Blanc wines display mineral aromas and flavours, vegetal character, or – in certain climates – fruity character, such as ripe melon, figs, or passion fruit. The wines are light- to medium-bodied and usually dry. Most of them are unoaked, but some are oaky.

France has two classic wine regions for the Sauvignon Blanc grape: Bordeaux and the Loire Valley, where the two best known Sauvignon wines are called Sancerre or Pouilly-Fumé.

Other white grapes

The following list describes some other grapes whose names you see on wine labels or whose wine you could drink in place-name wines without realizing it.

Albariño: An aromatic grape from the northwestern corner of Spain – the region called Rias Baixas – and Portugal's northerly Vinho Verde region, where it's called Alvarinho. It makes medium-bodied, crisp, appley-tasting, usually unoaked white wines whose high glycerin gives them silky texture.

Chenin Blanc: A noble grape in the Loire Valley of France, for Vouvray and other wines. The best wines have high acidity and a fascinating oily texture. Some good dry Chenin Blanc comes from California, but so does a ton of ordinary off-dry wine.

Gewürztraminer: A wonderfully exotic grape that makes fairly deep-coloured, full-bodied, soft white wines with aromas and flavours of roses and lychee fruit. France's Alsace region is the classic domain of this variety; the wines have pronounced floral and fruity aromas and flavours, but are actually dry – as fascinating as they are delicious.

Grüner Veltliner: A native Austrian variety that boasts complex aromas and flavours (vegetal, spicy, mineral), rich texture, and usually substantial weight.

Muscat: An aromatic grape that makes Italy's sparkling Asti (which, incidentally, tastes exactly like ripe Muscat grapes). It has extremely pretty floral aromas. In Alsace and Austria, Muscat makes a dry wine, and in lots of places (southern France, southern Italy, Australia), it makes a delicious, sweet dessert wine through the addition of alcohol.

Pinot Blanc: Fairly neutral in aroma and flavours, yet can make characterful wines. High acidity and low sugar levels translate into dry, crisp, medium-bodied wines. Alsace, Austria, northern Italy, and Germany are the main production zones.

Sémillon: Sauvignon Blanc's classic blending partner and a good grape in its own right. Sémillon wine is low in acid relative to Sauvignon Blanc and has attractive but subtle aromas – lanolin sometimes, although it can be slightly herbaceous when young. Sémillon is a major grape in Australia, and southwestern France, including Bordeaux (where it is the key player in the dessert wine, Sauternes).

Viognier: A grape from France's Rhône Valley that's becoming popular in California, the south of France and elsewhere. It has a floral aroma, delicately apricot-like, medium- to full-bodied with low acidity.

Pinot Gris/Pinot Grigio

Pinot Gris (*gree*), which is called *Pinot Grigio* in Italian, is one of several grape varieties called *Pinot*. Pinot Gris is believed to have mutated from the black Pinot Noir grape. Although it's considered a white grape, its skin colour is unusually dark for a white variety.

Wines made from Pinot Gris can be deeper in colour than most white wines – although most of Italy's Pinot Grigio wines are quite pale. Pinot Gris wines are medium- to full-bodied, usually not oaky, and have rather low acidity and fairly neutral aromas. Sometimes the flavour and aroma can suggest the skins of fruit, such as peach skins or orange rind.

Pinot Gris is an important grape throughout Northeastern Italy and also grows in Germany. The only region in France where Pinot Gris is important is in Alsace, where it really struts its stuff.

A Primer on Red Grape Varieties

Here are descriptions of 12 important red vinifera grape varieties. You'll encounter these grapes in varietal wines and also in place-name wines.

Cabernet Sauvignon

Cabernet Sauvignon is a noble grape variety that grows well in just about any climate that isn't very cool. It became famous through the age-worthy red wines of the Médoc district of Bordeaux (which usually also contain Merlot and Cabernet Franc, in varying proportions). But today California is an equally important region for Cabernet Sauvignon – not to mention Washington, southern France, Italy, Australia, South Africa, Chile, Argentina, and so on.

Because Cabernet Sauvignon is fairly tannic (and because of the blending precedent in Bordeaux), winemakers often blend it with other grapes; usually Merlot – being less tannic – is considered an ideal partner.

TIP

Cabernet Sauvignon wines come in all price and quality levels. The least expensive versions are usually fairly soft and very fruity, with medium body. The best wines are rich and firm with great depth and classic Cabernet flavour. Serious Cabernet Sauvignons can age for 15 years or more.

Merlot

Deep colour, full body, high alcohol, and low tannin are the characteristics of wines made from the Merlot grape. The aromas and flavours can be plummy or sometimes chocolatey, or they can suggest tea leaves.

TIP

Some wine drinkers find Merlot easier to like than Cabernet Sauvignon because it's less tannic. Merlot makes both inexpensive, simple wines and, when grown in the right conditions, very serious wines.

Pinot Noir

Pinot Noir is finicky, troublesome, enigmatic, and challenging to make. But a great Pinot Noir can be one of the greatest wines ever.

Pinot Noir wine is lighter in colour than Cabernet or Merlot. It has relatively high alcohol, medium-to-high acidity, and medium-to-low tannin (although oak barrels can contribute additional tannin to the wine). Its flavours and aromas can be very fruity – often like a mélange of red berries – or earthy and woodsy, depending on how it is grown and/or vinified. Pinot Noir is rarely blended with other grapes.

Syrah/Shiraz

The northern part of France's Rhône Valley is the classic home for great wines from the Syrah grape. Syrah produces deeply coloured wines with full body, firm tannin, and aromas/flavours that can suggest berries, smoked meat, black pepper, tar, or even burnt rubber (believe it or not). Syrah doesn't require any other grape to complement its flavours, although in Australia, it's often blended with Cabernet, and in the Southern Rhône, it's often part of a blended wine with Grenache and other varieties.

Zinfandel

White Zinfandel is such a popular wine – and so much better known than the red style of Zinfandel – that its fans might argue that Zinfandel is a white grape. But it's really red.

Zin makes rich, dark wines that are high in alcohol and medium to high in tannin. They can have a blackberry or raspberry aroma and flavour, a spicy or tarry character, or even a jammy flavour. Some Zins are lighter than others and meant to be enjoyed young, and some are serious wines with a tannin structure that's built for aging. (You can tell which is which by the price.)

Nebbiolo

Outside of scattered sites in Northwestern Italy – mainly the Piedmont region – Nebbiolo just doesn't make remarkable wine. But the extraordinary quality of Barolo and Barbaresco, two Piedmont wines, prove what greatness it can achieve under the right conditions.

The Nebbiolo grape is high in both tannin and acid, which can make a wine tough. Fortunately, it also gives enough alcohol to soften the package. Its colour can be deep when the wine is young but can develop orangey tinges within a few years. Its complex aroma is fruity (strawberry, cherry), earthy and woodsy (tar, truffles), herbal (mint, eucalyptus, anise), and floral (roses).

Lighter versions of Nebbiolo are meant to be drunk young – wines labeled Nebbiolo d'Alba, Roero, or Nebbiolo delle Langhe, for example – while Barolo and Barbaresco are wines that really deserve a *minimum* of eight years' age before drinking.

Sangiovese

This Italian grape has proven itself in the Tuscany region of Italy, especially in the Brunello di Montalcino and Chianti districts. Sangiovese makes wines that are medium to high in acidity and firm in tannin; the wines can be light-bodied to full-bodied, depending on exactly where the grapes grow and how the wine is made. The aromas and flavours of the wines are fruity – especially cherry, often tart cherry – with floral nuances of violets and sometimes a slightly nutty character.

Tempranillo

Tempranillo is Spain's candidate for greatness. It gives wines deep colour, low acidity, and only moderate alcohol. Modern renditions of Tempranillo from the Ribera del Duero region and elsewhere in Spain prove what colour and fruitiness this grape has. In more traditional wines, such as those of the Rioja region, much of the grape's colour and flavour is lost due to long wood aging and to blending with varieties that lack colour, such as Grenache.

Other red grapes

The following list describes additional red grape varieties and their wines, which you can encounter either as varietal wines or as wines named for their place of production.

Aglianico: From Southern Italy, where it makes Taurasi and other age-worthy, powerful red wines, high in tannin.

Barbera: Italian variety that, oddly for a red grape, has little tannin but very high acidity. When fully ripe, it can give big, fruity wines with refreshing crispness. Many producers age the wine in new oak to increase the tannin level of their wine.

Cabernet Franc: A parent of Cabernet Sauvignon, and often blended with it to make Bordeaux-style wines. It ripens earlier and has more expressive, fruitier flavour (especially berries), as well as less tannin.

Gamay: Excels in the Beaujolais district of France. It makes grapey wines that can be low in tannin – although the grape itself is fairly tannic.

Grenache: A Spanish grape by origin, called Garnacha there. (Most wine drinkers associate Grenache with France's southern Rhône Valley more than with Spain, however.) Sometimes Grenache makes pale, high-alcohol wines that are dilute in flavour. In the right circumstances, it can make deeply coloured wines with velvety texture and fruity aromas and flavours suggestive of raspberries.

Judging a Wine by Its Label

IN THIS ARTICLE

- *Deciphering the language of the wine label*
- *Figuring out place-names*
- *Discovering the difference between vintage and nonvintage wines*

Do you sometimes feel at a loss when you stand in front of an array of wine bottles and attempt to make a judgment about which to buy? To the untrained eye, wine labels hardly say anything, and the pretty pictures on the labels are even less relevant to what's inside the bottle.

In this article, you discover how to gather the information you really need from that confusing wine label. Before you know it, you'll be deciphering labels like a pro.

The Forward and Backward of Wine Labels

Many wine bottles have two labels. The front label names the wine and grabs your eye as you walk down the aisle, and the back label gives you a little more information, ranging from really helpful suggestions like "this wine tastes delicious with food" to useful data such as "this wine has a total acidity of 6.02 and a pH of 3.34."

TIP

If you're really on your toes, you may be thinking: How can you tell the difference between front and back on a round bottle? Governments require certain information to appear on the front label of all wine bottles – basic stuff, such as the alcohol content, the type of wine (usually *red table wine* or *white table wine*), and the country of origin – but they don't define *front label.* So sometimes producers put all that information on the smaller of two labels and call that one the front label. Then the producers place a larger, colourful, dramatically eye-catching label – with little more than the name of the wine on it – on the *back* of the bottle. Guess which way the back label ends up facing when the bottle is placed on the shelf?

Labeling: The Mandatory Sentence

Governments often stipulate that certain items of information appear on labels of wines sold in the United States. Such items are generally referred to as the mandatory. These include

- A brand name
- Indication of class or type (table wine, dessert wine, or sparkling wine)
- The percentage of alcohol by volume (unless it is implicit – for example, the statement "table wine" implies an alcohol content of less than 14 per cent)
- Name and location of the bottler
- Net contents (expressed in milliliters; the standard wine bottle is 750 ml)
- The phrase *Contains Sulphites* (with very, very few exceptions)
- The government warning

Wines made outside the United States but sold within it must also carry the phrase *imported by* on their labels, along with the name and business location of the importer.

Canadian regulations are similar. They require wine labels to indicate the *common name* of the product (that is, *wine*), the net contents, the per centage of alcohol by volume, the name and address of the producer, the wine's country of origin, and the container size.

Some of the mandatory information on American and Canadian wine labels is also required by the EU authorities for wines produced or sold in the European Union. But the EU regulations require additional label items for wines produced in its member countries.

The most important of these additional items is an indication of a wine's so-called quality level – which really means the wine's status in the European Union's hierarchy of place-names. In short, every wine made in an EU member country *must* carry one of the following items on the label:

- A registered place-name, along with an official phrase that confirms that the name is in fact a registered place-name
- A phrase indicating that the wine is a table wine, a status lower than that of a wine with a registered place-name

Appellations of Origin

A registered place-name is called an *appellation of origin.* In actuality, each EU place-name defines far more than just the name of the place that the grapes come from. The place-name connotes the wine's grape varieties, grape-growing methods, and winemaking methods. Each appellation is, therefore, a definition of the wine as well as the wine's name.

Wines with official place-names

European wines with official place-names fall into a European category called QWPSR (Quality Wine Produced in a Specific Region). The following phrases on European labels confirm that a wine is a QWPSR wine and that its name is therefore a registered place-name:

France: *Appellation Contrôlée* or *Appellation d'Origine Contrôlée* (AC or AOC, in short), translated as *regulated name or regulated place-name.* Also, on labels of wines from places of slightly lower status, the initials AO VDQS, standing for *Appellation d'Origine – Vins Délimités de Qualité Supérieure;* translated as *place-name, demarcated wine of superior quality.*

Italy: *Denominazione di Origine Controllata* (DOC), translated as *regulated place-name;* or for certain wines of an even higher status, *Denominazione di Origine Controllata e Garantita* (DOCG), translated as *regulated and guaranteed place-name.*

Spain: *Denominación de Origen* (DO), translated as *place-name;* and *Denominación de Origen Calificada* (DOC), translated as *qualified-origin place-name* for regions with the highest status (of which there are only two, Rioja and Priorat).

Portugal: *Denominação de Origem* (DO), translated as *place-name.*

Germany: *Qualitätswein bestimmter Anbaugebiete* (QbA), translated as *quality wine from a specific region;* or *Qualitätswein mit Prädikat* (QmP), translated as *quality wine with special attributes,* for the best wines.

Wines without an official appellation of origin

For European table wines — wines without an official appellation of origin — each European country has two phrases. One term applies to table wines with a geographic indication (actually, Italy has two phrases in this category), and another denotes table wines with no geographic indication smaller than the country of production. These phrases are

France: Vin de pays (country wine) followed by the name of an approved area; vin de table

Italy: Indicazione Geografica Tipica (translated as typical geographic indication and abbreviated as IGT) and the name of an approved area, or vino da tavola (table wine) followed by a the name of a geographic area; vino da tavola

Spain: Vino de la tierra (country wine) followed by the name of an approved area; vino de mesa

Portugal: Vinho Regional (regional wine) and the name of an approved area; vinho de mesa

Germany: Landwein (country wine) and the name of an approved area; Deutscher tafelwein

European wine designations at a glance

The following table highlights the European wine designations for easy reference:

Country	QWPSR Designation(s)	Table Wine Designation with Geographic Indication	Table Wine Designation without Geographic Indication
France	AOC VDQS	Vin de pays	Vin de table
Italy	DOCG DOC	IGT; Vino da tavola (and geographic name)	Vino da tavola
Spain	DOC DO	Vino de la tierra	Vino de mesa
Portugal	DO	Vinho regional	Vinho de mesa
Germany	QmP QbA	Landwein	Deutscher tafelwein

Some Optional Label Lingo

Besides the mandatory information required by government authorities, all sorts of other words can appear on wine labels. These words can be meaningless phrases intended to make you think that you're getting a special quality wine or words that provide useful information about what's in the bottle. Sometimes the same word can fall into either category, depending on the label. This ambiguity occurs because some words that are strictly regulated in some producing countries aren't at all regulated in others.

Vintage

The word *vintage* followed by a year, or the year listed alone without the word *vintage,* is the most common optional item on a wine label. Sometimes the vintage appears on the front label, and sometimes it has its own small label above the front label.

Generally speaking, *what* vintage a wine is – that is, whether the grapes grew in a year with perfect weather or whether the grapes were meteorologically challenged – is an issue you need to consider a) only when you buy top quality wines, and b) mainly when those wines come from parts of the world that experience significant variations in weather from year to year – such as many European wine regions.

REMEMBER

The *vintage year* is nothing more than the year in which the grapes for a particular wine grew; the wine must have 75 to 100 per cent of the grapes of this year, depending on the country of origin. (*Non-vintage* wines are blends of wines whose grapes were harvested in different years.) But there is an aura surrounding vintage-dated wine that causes many people to believe that any wine with a vintage date is by definition better than a wine without a vintage date. *In fact, there is no correlation between the presence of a vintage date and the wine's quality.*

The EU hierarchy of wine

Although each country within the European Union makes its own laws regarding the naming of wine, these laws must fit within the framework of the European Union law. This framework provides two levels into which every single EU-produced wine must fit:

- **Quality wine:** Wines with official appellations of origin. (Each appellation regulation defines the geographic area, the grapes that may be used, grape-growing practices, winemaking and aging techniques, and so on.) This category is abbreviated as QWPSR in English (Quality Wine Produced in a Specific Region) or VQPRD in several other European languages. All AOC, DOC, DO, and QbA wines – to use the abbreviations mentioned previously in this chapter – fall into this category.
- **Table wine:** All other wines produced within the European Union. The table wine category has two subcategories: Table wines that carry a precise geographic indication on their labels, such as French *vin de pays* or Spanish *vino de la tierra* wines; and table wines with no geographic indication except the country of origin. These latter wines may not carry a vintage or a grape name.

All other wines sold in the European Union fall into a third category: *wine*, which includes wines produced by countries outside the European Union, such as the United States, Canada, or Australia. If a wine has a geographic indication smaller than the country of origin, it enjoys higher status than otherwise.

Reserve

Reserve is a favourite word on wine labels. The term is used to convince you that the wine inside the bottle is special. This trick usually works because the word *does* have specific meaning and *does* carry a certain amount of prestige on labels of wines from many other countries:

- In Italy and Spain, the word *reserve* (or its foreign language equivalent, which looks something like *reserve*) indicates a wine that has received extra aging at the winery before release. Implicit in the extra aging is the idea that the wine was better than normal and, therefore, worthy of the extra aging. Spain even has *degrees* of reserve, such as Gran Reserva.
- In France, the use of *reserve* isn't regulated. However, its use is generally consistent with the notion that the wine is better in quality than a given producer's norm.

Estate-bottled

Estate is a genteel word for a wine farm, a combined grape-growing and winemaking operation. The words *estate-bottled* on a wine label indicate that the company that bottled the wine also grew the grapes and made the wine. In other words, *estate-bottled* suggests accountability from the vineyard to the winemaking through to the bottling. In many countries, the winery doesn't necessarily have to own the vineyards, but it has to control the vineyards and perform the vineyard operations.

TIP

Sometimes French wine labels carry the words *domaine-bottled* or *château-bottled* (or the phrase *mis en bouteille au château/au domaine*). The concept is the same as estate-bottled, with *domaine* and *château* being equivalent to the term *estate*.

Estate-bottling is an important concept to those who believe that you can't make good wine unless the grapes are as good as they can possibly be. Wine lovers wouldn't go so far as to say that great wines *must* be estate-bottled, though.

Vineyard name

Some wines in the medium-to-expensive price category – costing about £15 or more – may carry on the label the name of the specific vineyard where the grapes for that wine grew. Sometimes one winery will make two or three different wines that are distinguishable only by the vineyard name on the label. Each wine is unique because the *terroir* of each vineyard is unique. See article 4 for more on terroir. These single vineyards may or may not be identified by the word *vineyard* next to the name of the vineyard.

Italian wines, which are really into the single-vineyard game, will have *vigneto* or *vigna* on their labels next to the name of the single vineyard. Or they won't. It's optional.

Other optional words on the label

One additional expression on some French labels is *Vieilles Vignes (vee yay veen),* which translates as *old vines,* and appears as such on some Californian and Australian labels. Because old vines produce a very small quantity of fruit compared to younger vines, the quality of their grapes and of the resulting wine is considered to be very good. The problem is, the phrase is unregulated. Anyone can claim that his vines are old.

The word *superior* can appear in French *(Supérieure)* or Italian *(Superiore)* as part of an AOC or DOC place-name. It means the wine attained a higher alcohol level than a nonsuperior version of the same wine would have. Frankly, it's a distinction not worth losing sleep over.

The word *Classico* appears on the labels of some Italian DOC and DOCG wines when the grapes come from the heartland of the named place.

What grape variety made this wine?

Most New World wines (from the Americas, Australia, and other continents besides Europe) tell you what grape variety they're made from right on the front label – it's often the very name of the wine – or on the back label. Traditional European wines blended from several grape varieties usually don't give you that information a) because the winemakers consider the name of the place more important than the grapes, anyway, and b) because often the grapes they use are local varieties whose names few people would recognize.

If you really want to know what grape varieties make a Soave, Valpolicella, Châteauneuf-du-Pape, Rioja, Côtes du Rhône, or other blended European wines, you'll generally have to look it up yourself.

The Wine Name Game

IN THIS ARTICLE

- *Decoding wine names*
- *Getting the scoop on the secret cult of terroir*
- *Exploring branded and generic wines*

Never before have wine shops offered such an astounding proliferation of wine labels! Since about 2003, it seems that new brands of wine have appeared out of the blue every week.

All this choice is terrific — or it's completely paralyzing, depending on how you approach the situation. More than half the white wines are named Chardonnay, and the majority of the red wines are named either Cabernet Sauvignon or Merlot. But one Merlot isn't identical to another Merlot.

What's in a Name

All sorts of names appear on wine labels. These names often include

- The name of the grape from which the wine was made.
- A brand name, which is traditionally the name of the company or person that made the wine (who is called the *producer*), but for less expensive wines is likely to be an invented name.
- Sometimes a special, fanciful name for that particular wine (called a *proprietary name*).
- The name of the place, or places, where the grapes grew (the wine region, and sometimes the name of the specific vineyard property).

Then there's the vintage year (the year the grapes for that wine grew), which is part of the wine's identity; and sometimes you see a descriptor like reserve, which either has specific legal meaning or means nothing at all, depending on where the wine came from.

REMEMBER

Veteran lovers appreciate all this detailed information on wine labels because they know how to interpret it. But to anyone who is just discovering wine, the information embedded in wine names is more confusing than enlightening. Do you really need to know all these details? For bottles of wine, the answer is yes.

Naming Wines: Is It a Grape? Is It a Place?

Most of the wines that you find in your wine shop or on restaurant wine lists are named in one of two basic ways:

- For their grape variety
- For the place where the grapes grew

That information, plus the name of the producer, becomes the shorthand name used when talking about the wine.

Robert Mondavi Cabernet Sauvignon, for example, is a wine made by Robert Mondavi Winery and named after the Cabernet Sauvignon grape. Fontodi Chianti Classico is a wine made by the Fontodi winery and named after the place called Chianti Classico.

You may recognize some names as grape names and other names as place-names right off the bat; but if you don't, don't panic. That information is the kind of thing you can look up.

It's a grape! Hello, my name is Chardonnay

A *varietal* wine is a wine that is named after either the *principal* or the *sole* grape variety that makes up the wine.

Each country (and in the United States, some individual states) has laws that dictate the minimum percentage of the named grape that a wine must contain if that wine wants to call itself by a grape name. The issue is truth in advertising.

Some varietal wines are made *entirely* from the grape variety for which the wine is named. There's no law against that anywhere.

WARNING!

Most of the time, the labels of varietal wines don't tell you whether other grapes are present in the wine, what those grapes are, or the percentage of the wine that they account for. All you know is that the wine contains at least the minimum legal percentage of the named variety.

Why name a wine after a grape variety? Grapes are the raw material of a wine. Except for whatever a wine absorbs from oak barrels (certain aromas and flavours, as well as tannin) and from certain winemaking processes, the juice of the grapes is what any wine *is.* So to name a wine after its grape variety is very logical.

Naming a wine for its grape variety is also very satisfying to exacting consumers. Knowing what grape a wine is made from is akin to knowing what type of oil is in the salad dressing, whether your bread contains any trans-fats, and how much egg is in your egg roll.

Most American wines carry varietal names. Likewise, most Australian, South American, and South African wines are named by using the *principal* principle. Even some countries that don't normally name their wines after grapes, such as France, are jumping on the varietal-name bandwagon for certain wines that they especially want to sell to Americans.

WARNING!

A common perception among some wine drinkers is that a varietal wine is somehow *better* than a non-varietal wine. Actually, the fact that a wine is named after its principal grape variety is absolutely *no indication of quality.*

It's a place! Hello, my name is Bordeaux

Most European wines are named for the *region* where their grapes grow rather than for the grape variety itself. Many of these European wines come from precisely the same grape varieties as American wines (like Chardonnay, Cabernet Sauvignon, Sauvignon Blanc, and so on), but they don't say so on the label. Instead, the labels say Burgundy, Bordeaux, Sancerre, and so on: the *place* where those grapes grow.

The European system of naming wines is actually intended to provide more information about each wine, and more understanding of what's in the bottle, than varietal naming does. The only catch is that to harvest this information, you have to learn something about the different regions from which the wines come.

The most common place-names are

Beaujolais	Chianti	Rioja
Bordeaux	Côtes du Rhône	Sancerre
Burgundy (Bourgogne)	Mosel	Sauternes
Chablis	Port (Porto)	Sherry
Champagne	Pouilly-Fuissé	Soave
Châteauneuf-du-Pape	Rhine (Rheingau, Rheinhessen)	Valpolicella

Why name a wine after a place?

Grapes, the raw material of wine, have to grow somewhere. Depending on the type of soil, the amount of sunshine, the amount of rain, the slope of the hill, and the many other characteristics that each somewhere has, the grapes will turn out differently. If the grapes are different, the wine is different. Each wine, therefore, reflects the place where its grapes grow.

In Europe, grape growers/winemakers have had centuries to figure out which grapes grow best where. They've identified most of these grape–location match-ups and codified them into regulations. Therefore, the name of a place where grapes are grown in Europe automatically connotes the grape (or grapes) used to make the wine of that place. The label on the bottle usually doesn't tell you the grape (or grapes), though.

The following table decodes common European place-names

Wine Name	Country	Grape Varieties
Beaujolais	France	Gamay
Bordeaux (red)	France	Cabernet Sauvignon, Merlot, Cabernet Franc, and others*
Bordeaux (white)	France	Sauvignon Blanc, Sémillon Muscadelle*
Burgundy (red)	France	Pinot Noir
Burgundy (white)	France	Chardonnay
Chablis	France	Chardonnay
Champagne	France	Chardonnay, Pinot Noir, Pinot Meunier*
Châteauneuf-du-Pape*	France	Grenache, Mourvèdre, Syrah, and others*
Chianti	Italy	Sangiovese, Canaiolo, and others*
Côtes du Rhône*	France	Grenache, Mourvèdre, Carignan, and others*
Port (Porto)	Portugal	Touriga Nacional, Tinta Barroca, Touriga Franca, Tinta Roriz, Tinto Cão and others*
Pouilly-Fuissé, Macon,	France	Chardonnay Saint Véran
Rioja (red)	Spain	Tempranillo, Grenache, and others*
Sancerre/Pouilly-Fumé	France	Sauvignon Blanc
Sherry	Spain	Palomino
Soave	Italy	Garganega and others*
Valpolicella	Italy	Corvina, Molinara, Rondinella*

*Indicates that a blend of grapes is used to make these wines.

Say What? The Terroir Game

Terroir (pronounced *ter wahr*) is a French word that has no direct translation in English, so wine people just use the French word, for expediency (not for snobbery). There's no fixed definition of *terroir;* it's a concept, and, like most concepts, people tend to define it more broadly or more narrowly to suit their own needs. The word itself is based on the French word *terre,* which means soil; so some people define *terroir* as, simply, dirt.

But *terroir* is really much more complex (and complicated) than just dirt. *Terroir* is the combination of immutable natural factors – such as topsoil, subsoil, climate (sun, rain, wind, and so on), the slope of the hill, and altitude – that a particular vineyard site has. Chances are that no two vineyards in the entire world have precisely the same combination of these factors. So *terroir* is the unique combination of natural factors that a particular vineyard site has.

Terroir is the guiding principle behind the European concept that wines should be named after the place they come from. The thinking goes like this: The name of the place connotes which grapes were used to make the wine of that place (because the grapes are dictated by regulations), and the place influences the character of those grapes in its own unique way. Therefore, the most accurate name that a wine can have is the name of the place where its grapes grew.

It's not some nefarious plot; it's just a whole different way of looking at things.

Wines Named in Other Ways

Now and then, you may come across a wine that is named for neither its grape variety nor its region of origin. Such wines usually fall into three categories: *branded wines, wines with proprietary names,* or *generic wines.*

Grape names on European wines

Although most European wines are named after their place of origin, grape names do sometimes appear on labels of European wines. In Italy, for example, the official name of a wine could be a combination of place and grape – like the name Barbera d'Alba, which translates as Barbera (grape) of Alba (place).

In France, some producers have deliberately added the grape name to their labels even though the grape is already implicit in the wine name. For example, a white Bourgogne (place-name) may also have the word Chardonnay (grape) on the label, for those wine drinkers who don't know that white Bourgogne is 100 percent Chardonnay. And German wines usually carry grape names along with their official place-names.

But even if a European wine does carry a grape name, the most important part of the wine's name, in the eyes of the people who make the wine, is the place.

Branded wines

Most wines have brand names, including those wines that are named after their grape variety – like Cakebread (brand name) Sauvignon Blanc (grape) – and those that are named after their region of origin, such as Masi (brand name) Valpolicella (place). These brand names are usually the name of the company that made the wine, called a *winery*. Because most wineries make several different wines, the brand name itself is not specific enough to be the actual name of the wine.

But sometimes a wine has *only* a brand name. For example, the label says *Salamandre* and *red French wine* but provides little other identification.

WARNING!

Wines that have *only* a brand name on them, with no indication of grape or of place – other than the country of production – are generally the most inexpensive, ordinary wines you can get. If they're from a European Union country, they won't even be *vintage dated* (that is, there won't be any indication of what year the grapes were harvested) because EU law does not entitle such wines to carry a vintage date.

Wines with proprietary names

You can find some pretty creative names on wine bottles these days: Tapestry, Conundrum, Insignia, Isosceles, Mythology, Trilogy. Names like these are *proprietary names* (often trademarked) that producers create for special wines. In the case of European wines, the grapes used to make the wine were probably not the approved grapes for that region; therefore, the regional name could not be used on the label. In the case of American wines, the bottles with proprietary names usually contain wines made from a *blend* of grapes; therefore, no one grape name can be used as the name of the wine.

Wines with proprietary names usually are made in small quantities, are quite expensive (£30 to £60 or more a bottle), and are high in quality.

REMEMBER

Although a brand name can apply to several different wines, a proprietary name usually applies to one specific wine. You can find Zinfandel, Cabernet Sauvignon, Chardonnay, and numerous other wines under the Fetzer brand from California, for example, and you can find Beaujolais, Pouilly-Fuissé, Mâcon-Villages, and numerous other wines under the Louis Jadot brand from France. But the proprietary name Luce applies to a single wine.

Generic wines

A generic name is a wine name that has been used inappropriately for so long that it has lost its original meaning in the eyes of the government (exactly what Xerox, Kleenex, and Band-Aid are afraid of becoming).

Burgundy, Chianti, Chablis, Champagne, Rhine wine, Sherry, Port, and Sauterne are all names that rightfully should apply only to wines made in specific places. After years of negotiation with the European Union, the U.S. government has finally agreed that these names can no longer be used for American wines. However, any wine that bore such a name prior to March 2006 may continue to carry that name. In time, generic names will become less common on wine labels.

These Taste Buds Are for You

IN THIS ARTICLE

- *Tasting wine the right way*
- *Judging wine by appearance*
- *Smelling aromas in wine*
- *Describing taste to others*

You drink beverages every day, tasting them as they pass through your mouth. In the case of wine, however, drinking and tasting are not synonymous. Wine is much more complex than other beverages: There's more going on in a mouthful of wine. Most wines have a lot of different (and subtle) flavours, all at the same time, and they give you multiple sensations when they're in your mouth, such as softness and sharpness together.

If you just drink wine, gulping it down the way you do soda, you miss a lot of what you paid for. But if you *taste* wine, you can discover its nuances. In fact, the more slowly and attentively you taste wine, the more interesting it tastes.

Tasting Wine: Look, Sniff, and Then Taste

The two fundamental rules of wine tasting are to slow down and pay attention. The process itself involves looking, smelling, and then tasting the wine.

REMEMBER

When it comes to smelling wine, many people are concerned that they aren't able to detect as many aromas as they think they should. Smelling wine is really just a matter of practice and attention. If you start to pay more attention to smells in your normal activities, you'll get better at smelling wine.

TIP

Savoring wine's good looks

To observe a wine's appearance, tilt a (half-full) glass away from you and look at the colour of the wine against a white background, such as the tablecloth or a piece of paper (a coloured background distorts the colour of the wine). Notice how dark or how pale the wine is, what colour it is, and whether the colour fades from the centre of the wine out toward the edge, where it touches the glass. Also notice whether the wine is cloudy, clear, or brilliant. (Most wines are clear. Some wines that are *unfiltered* can be less than brilliant, but they shouldn't be cloudy.) Eventually, you'll begin to notice patterns, such as deeper colour in younger red wines.

Swirling and sniffing: The nose knows

The swirling and sniffing step is when you can let your imagination run wild, and no one will ever dare to contradict you. If you say that a wine smells like wild strawberries to you, how can anyone prove that it doesn't?

Keep your glass on the table and rotate it three or four times so that the wine swirls around inside the glass and mixes with air. Then quickly bring the glass to your nose. Stick your nose into the airspace of the glass, and smell the wine. Free-associate. Is the aroma fruity, woodsy, fresh, cooked, intense, light? Your nose tires quickly, but it recovers quickly, too. Wait just a moment and try again. Listen to your friends' comments and try to find the same things they find in the smell.

As you swirl, the aromas in the wine vaporize so that you can smell them. Wine has so many *aromatic compounds* that whatever you find in the smell of a wine is probably not merely a figment of your imagination. Common aromas (or flavours) associated with wine include fruits, herbs, flowers, earth, grass, tobacco, butterscotch, toast, vanilla, and coffee, mocha, or chocolate.

These tips can help you detect aromas:

- **Be bold.** Stick your nose right into the airspace of the glass where the aromas are captured.
- **Don't wear a strong scent**; it will compete with the smell of the wine.
- **Don't knock yourself out smelling a wine when there are strong food aromas around.** The tomatoes you smell in the wine could really be the tomato in someone's pasta sauce.
- **Become a smeller.** Smell every ingredient when you cook, everything you eat, the fresh fruits and vegetables you buy at the supermarket, even the smells of your environment – like leather, flowers, or your wet dog. Stuff your mental database with smells so that you'll have aroma memories at your disposal when you need to draw on them.
- **Try different sniffing techniques.** Some people like to take short, quick sniffs, while others like to inhale a deep whiff of the wine's smell. Keeping your mouth open a bit while you inhale can help you perceive aromas.

WARNING!

> To get the most out of your sniffing, swirl the wine in the glass first. But don't even *think* about swirling your wine if your glass is more than half full.

Ewww, what's that smell?

The point behind the ritual of swirling and sniffing is that what you smell should be pleasurable to you, maybe even fascinating, and that you should have fun in the process. But what if you notice a smell that you don't like?

Hang around wine geeks for a while, and you'll start to hear words like *petrol, manure, sweaty saddle, burnt match,* and *asparagus* used to describe the aromas of some wines. Fortunately, the wines that exhibit such smells aren't the wines you'll be drinking for the most part – at least not unless you really catch the wine bug. And when you do catch the wine bug, you may discover that those aromas, in the right wine, can really be a kick. Even if you don't learn to enjoy those smells (some people do, honest!), you'll appreciate them as typical characteristics of certain regions or grapes.

Then there are the bad smells that nobody will try to defend. It doesn't happen often, but it does happen, because wine is a natural, agricultural product with a will of its own. Often when a wine is seriously flawed, it shows immediately in the nose of the wine. Wine judges have a term for such wines. They call them DNPIM – Do Not Put In Mouth. Not that you'll get ill, but why subject your taste buds to the same abuse that your nose just took? Sometimes it's a bad cork that's to blame, and sometimes it's some other sort of problem in the winemaking or even the storage of the wine. Just rack it up to experience and open a different bottle.

The mouth action

After you've looked at the wine and smelled it, you're finally allowed to taste it. Here's how the procedure goes.

1. **Take a medium-sized sip of wine.**
2. **Hold it in your mouth, purse your lips, and draw in some air across your tongue, over the wine.**

   Be careful not to choke or dribble, or everyone will strongly suspect that you're not a wine expert.
3. **Swish the wine around in your mouth as if you're chewing it and then swallow it.**

The whole process should take several seconds, depending on how much you're concentrating on the wine.

Feeling the tastes

Taste buds on the tongue can register various sensations, which are known as the basic tastes. These include sweetness, sourness, saltiness, bitterness, and umami, a savory characteristic. Of these tastes, sweetness, sourness and bitterness are those most commonly found in wine. By moving the wine around in your mouth, you give it a chance to hit all your taste buds so that you don't miss anything in the wine (even if sourness and bitterness sound like things you wouldn't mind missing).

As you swish the wine around in your mouth, you're also buying time. Your brain needs a few seconds to figure out what the tongue is tasting and make some sense of it. Any sweetness in the wine registers in your brain first because many of the taste buds on the front of your tongue–where the wine hits first – capture the sensation of sweetness; *acidity* (which, by the way, is what normal people call sourness) and bitterness register subsequently. While your brain is working out the relative impressions of sweetness, acidity, and bitterness, you can be thinking about how the wine feels in your mouth – whether it's heavy, light, smooth, rough, and so on.

Tasting the smells

Until you cut your nose in on the action, that's all you can taste in the wine – those three sensations of sweetness, acidity, and bitterness and a general impression of weight and texture.

Do you like what you tasted? The possible answers are yes, no, an indifferent shrug of the shoulders, or "I'm not sure, let me take another taste."

TIP

If you're tasting wild strawberries, chocolate, and plums, you're actually tasting aromas, not through tongue contact, but by inhaling them up an interior nasal passage in the back of your mouth called the *retronasal passage* (see figure). When you draw in air across the wine in your mouth, you're vaporizing the aromas just as you did when you swirled the wine in your glass.

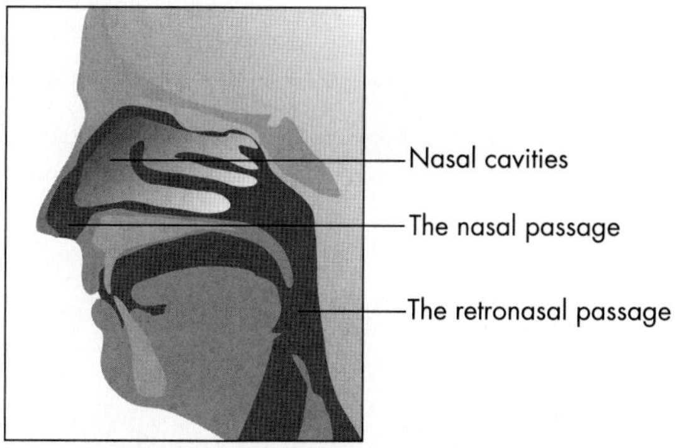

Wine flavours are actually aromas that vaporize in your mouth; you perceive them through the rear nasal passage.

Putting Taste into Words

"Like/Don't Like" is a no-brainer once you have the wine in your mouth. But most of the time, you have to buy the stuff without tasting it first. So unless you want to drink the same wine for the rest of your life, you're going to have to decide what it is that you like or don't like in a wine and communicate that to another person who can steer you towards a wine you'll like.

There are two hurdles to putting taste into words: Finding the words to describe what you like or don't like, and then getting the other person to understand what you mean.

The tastes of a wine reveal themselves sequentially as the tongue detects them, and they register in your brain. Follow this natural sequence when you try to put words to what you're tasting.

Taste for sweetness

As soon as you put the wine into your mouth, you can usually notice sweetness or the lack of it. In winespeak, *dry* is the opposite of sweet. Classify the wine you're tasting as either *dry, medium dry* (in other words, somewhat sweet), or *sweet.*

Sweetness, on the other hand, is a tactile impression on your tongue. When in doubt, try holding your nose when you taste the wine; if the wine really is sweet, you'll be able to taste the sweetness despite the fact that you can't smell the fruitiness.

TIP

Beginning wine tasters sometimes describe dry wines as sweet because they confuse fruitiness with sweetness. A wine is *fruity* when it has distinct aromas and flavours of fruit. You smell the fruitiness with your nose; in your mouth, you "smell" it through your retronasal passage.

Wines have noses – and palates, too

With poetic licence typical of wine tasters, someone once dubbed the smell of a wine its *nose* – and the expression took hold. If someone says that a wine has a huge nose, he means that the wine has a very strong smell. If he says that he detects lemon *in the nose* or *on the nose,* he means that the wine smells a bit like lemons.

In fact, most wine tasters rarely use the word *smell* to describe how a wine smells because the word *smell* (like the word *odour*) seems pejorative. Wine tasters talk about the wine's nose or aroma. Sometimes they use the word *bouquet,* although that word is falling out of fashion.

Just as a wine taster might use the term *nose* for the smell of a wine, he might use the word *palate* in referring to the taste of a wine. A wine's palate is the overall impression the wine gives in your mouth, or any isolated aspect of the wine's taste – as in "This wine has a harmonious palate," or "The palate of this wine is a bit acidic." When a wine taster says that he finds raspberries *on the palate,* he means that the wine has the flavour of raspberries.

Taste for acidity

All wine contains acid (mainly *tartaric acid,* which exists in grapes), but some wines are more acidic than others. Acidity is more of a taste factor in white wines than in reds. For white wines, acidity is the backbone of the wine's taste (it gives the wine firmness in your mouth). White wines with a high amount of acidity feel *crisp,* and those without enough acidity feel *flabby.*

You generally perceive acidity in the middle of your mouth – what wine-tasters call the *mid-palate.* You can also sense the consequences of acidity (or the lack of it) in the overall style of the wine – whether it's a tart little number or a soft and generous sort, for example. Classify the wine you're tasting as *crisp, soft,* or "couch potato."

Taste for tannin

Tannin is a substance that exists naturally in the skins, seeds (or *pips*), and stems of grapes. Because red wines are fermented with their grape skins and pips, and because red grape varieties are generally higher in tannin than white varieties, tannin levels are far higher in red wines than in white wines. Oak barrels can also contribute tannin to wines, both reds and whites.

To generalize a bit, tannin is to a red wine what acidity is to a white: a backbone. Tannins alone can taste bitter, but some tannins in wine are less bitter than others. Also, other elements of the wine, such as sweetness, can mask the perception of bitterness. You sense tannin – as bitterness, or as firmness or richness of texture – mainly in the rear of your mouth and, if the amount of tannin in a wine is high, on the inside of your cheeks and on your gums. Depending on the amount and nature of its tannin, you can describe a red wine as *astringent, firm,* or *soft.*

TIP

Have you ever taken a sip of a red wine and rapidly experienced a drying-out feeling in your mouth, as if something had blotted-up all your saliva? That's tannin.

TIP

Red wines have acid as well as tannin, and distinguishing between the two as you taste a wine can be a real challenge. When you're not sure whether it's mainly tannin or acid you're perceiving, pay attention to how your mouth feels *after* you've swallowed the wine. Acid makes you salivate (saliva is alkaline, and it flows to neutralize the acid). Tannin leaves your mouth dry.

Taste for body

A wine's body is an impression you get from the whole of the wine – not a basic taste that registers on your tongue. It's the impression of the weight and size of the wine in your mouth, which is usually attributable principally to a wine's alcohol. Obviously, one ounce of any wine will occupy exactly the same space in your mouth and weigh the same as one ounce of any other wine. But some wines *seem* fuller, bigger, or heavier in the mouth than others. Think about the wine's fullness and weight as you taste it. Imagine that your tongue is a tiny scale and judge how much the wine is weighing it down. Classify the wine as *light-bodied, medium-bodied,* or *full-bodied.*

The Flavour Dimension

Wines have flavours (er, *mouth aromas*), but wines don't come in a specific flavour. While you may enjoy the suggestion of chocolate in a red wine that you're tasting, you wouldn't want to go to a wine store and ask for a chocolaty wine.

If you like a wine and want to try another wine that's similar but different, one method is to decide what families of flavours in the wine you like and mention that to the person selling you your next bottle.

Another aspect of flavour that's important to consider is a wine's *flavour intensity* –how much flavour the wine has, regardless of what those flavours are. Flavour intensity is a major factor in pairing wine with food, and it also helps determine how much you like a wine.

TIP

Try to refer to *families of flavours* in wine. You have your *fruity wines* (the ones that make you think of all sorts of fruit when you smell them or taste them), your *earthy wines* (these make you think of minerals and rocks, walks in the forest, turning the earth in your garden, dry leaves, and so on), your *spicy wines* (cinnamon, cloves, black pepper, or Indian spices, for example), your *herbal wines* (mint, grass, hay, rosemary, and so on), and so on.

Finding and Buying Good Wine

IN THIS ARTICLE

- *Knowing what good wine tastes like*
- *Identifying bad wines*
- *Figuring out where to buy good wine*

Wine producers constantly brag about the quality ratings that their wines receive from critics, because a high rating — implying high quality — translates into increased sales for a wine. But quality wines come in all colours, degrees of sweetness and dryness, and flavour profiles.

REMEMBER

Just because a wine is high quality doesn't mean that you will actually enjoy it, any more than five star review means that you'll love a particular movie. You can still purchase a highly rated wine and end up pouring it down the sink because you didn't care to drink it. Personal taste is simply more relevant than quality in choosing a wine.

What's a Good Wine?

A good wine is, above all, a wine that you like enough to drink — because the whole purpose of a wine is to give pleasure to those who drink it. After that, how good a wine is depends on how it measures up to a set of (more or less) agreed-upon standards of performance established by experienced, trained experts. These standards involve mysterious concepts like *balance, length, depth, complexity,* and *trueness to type (typicity).* None of these concepts is objectively measurable, by the way.

TIP

The instruments that measure the quality of a wine are a human being's nose, mouth, and brain. Because everyone's different, everyone has different opinions on how good a wine is. The combined opinion of a group of trained, experienced palates (also known as wine experts) is usually considered a definitive judgement of a wine's quality.

Does a wine have to be expensive to be good?

For wine, as for many other products, a high price often indicates high quality. But the highest quality wine isn't always the best choice, for the following reasons:

- Your taste is personal, and you may not like a wine that critics consider very high in quality.
- Not all situations call for a very high quality wine.

You can enjoy even a £5 wine in many circumstances. At large family gatherings, on picnics, at the beach, and so on, an expensive, top-quality wine can be out-of-place – too serious and important.

Likewise, the very finest wines are seldom the best choices in restaurants – considering typical restaurant prices. Instead, look for the best value on the wine list (keeping in mind what you're eating) or experiment with some moderately priced wine that you haven't tried before. (There will always be some wines that you haven't tried.)

Quality isn't the only consideration in choosing a wine. Often, the best wine for your taste or for a certain situation will be inexpensive.

Balance

Three words – sweetness, acidity, and tannin – represent three of the major *components* (parts) of wine. The fourth is alcohol. Besides being one of the reasons you usually want to drink a glass of wine in the first place, alcohol is an important element of wine quality.

Balance is the relationship of these four components to one another. A wine is balanced when nothing sticks out as you taste it, such as harsh tannin or too much sweetness. Most wines are balanced to most people. But if you have any pet peeves about food – if you really hate anything tart, for example, or if you never eat sweets – you may perceive some wines to be unbalanced. If you perceive them to be unbalanced, then they're unbalanced for you. (Professional tasters know their own idiosyncrasies and adjust for them when they judge wine.)

Tannin and acidity are *hardening elements* in a wine (they make a wine taste firmer in the mouth), while alcohol and sugar (if any) are *softening elements*. The balance of a wine is the interrelationship of the hard and the soft aspects of a wine, and a key indicator of quality.

Balance in action

For a firsthand experience of how the principle of taste balance works, try this. Make a very strong cup of tea and chill it. When you sip it, the cold tea will taste bitter because it's very tannic. Now add lemon juice; the tea will taste astringent (constricting the pores in your mouth) because the acid of the lemon and the tannin of the tea are accentuating each other. Now add a lot of sugar to the tea. The sweetness should counterbalance the acid–tannin impact, and the tea will taste softer than it did before.

Length

When wines are called *long* or *short*, it's not referring to the size of the bottle or how quickly you empty it. *Length* is a word used to describe a wine that gives an impression of going all the way on the palate – you can taste it across the full length of your tongue – rather than stopping short halfway through your tasting of it.

Many wines today are very up front on the palate – they make a big impression as soon as you taste them – but they don't go the distance in your mouth. They are *short*. Generally, high alcohol or excess tannin is to blame. Length is a sure sign of high quality.

Depth

This is another subjective, unmeasurable attribute of a high-quality wine. A wine has *depth* when it seems to have a dimension of verticality — that is, it doesn't taste flat and one-dimensional in your mouth. A flat wine can never be great.

Complexity

Nothing's wrong with a simple, straightforward wine, especially if you enjoy it. But a wine that keeps revealing different things about itself, always showing you a new flavour or impression – a wine that has *complexity* – is usually considered better quality. Some experts use the term *complexity* specifically to indicate that a wine has a multiplicity of aromas and flavours, while others use it in a more holistic (but less precise) sense, to refer to the total impression a wine gives you.

Finish

The impression a wine leaves in the back of your mouth and in your throat after you've swallowed it is its *finish* or *aftertaste.* In a good wine, you can still perceive the wine's flavours – such as fruitiness or spiciness – at that point. Some wines may finish *hot,* because of high alcohol, or *bitter,* because of tannin – both shortcomings. Or a wine may have nothing much at all to say for itself after you swallow.

Typicity

To judge whether a wine is true to its type, you have to know how that type is supposed to taste. So you have to know the textbook characteristics of wines made from the major grape varieties and wines of the world's classic wine regions. (For example, the Cabernet Sauvignon grape typically has an aroma and flavour of blackcurrants, and the French white wine called Pouilly-Fumé typically has a slight gunflint aroma.)

REMEMBER

If you find a bad wine or a bad bottle – or even a wine that is considered a good wine, but you don't like it – just move on to something you like better.

Drinking a so-called great wine that you don't enjoy is as stupid as watching a television show that bores you. Change the channel. Explore.

What's a Bad Wine?

Strangely enough, the right to declare a wine "good" because you like it doesn't carry with it the right to call a wine "bad" just because you don't. In this game, you get to make your own rules, but you don't get to force other people to live by them.

And to be honest, very few bad wines are in the world. And many of the wines you could call bad are actually just bad *bottles* of wine – bottles that were handled badly so that the good wine inside them got ruined.

Here are some characteristics that everyone agrees indicate a bad wine:

- **Mouldy fruit:** Have you ever eaten a raspberry from the bottom of the container that had a dusty, cardboardy taste to it? That same taste of rot can be in a wine if the wine was made from grapes that weren't completely fresh and healthy when they were harvested. Bad wine.
- **Vinegar:** In the natural evolution of things, wine is just a passing stage between grape juice and vinegar. Most wines today remain forever in the wine stage because of technology or careful winemaking. If you find a wine that has crossed the line toward vinegar, it's bad wine.
- **Chemical or bacterial smells:** The most common odours are acetone (nail polish thinner) and sulphur flaws (rotten eggs, burnt rubber, bad garlic). Bad wines.
- **Oxidized wine:** This wine smells flat, weak, or maybe cooked, and it tastes the same. It may have been a good wine once, but air – oxygen – got in somehow and killed the wine. Bad bottle.
- **Cooked aromas and taste:** When a wine has been stored or shipped in heat, it can actually taste cooked or baked as a result. Often, you see telltale leakage from the cork, or the cork has pushed up a bit. Bad bottle. (Unfortunately, every other bottle of that wine that experienced the same shipping or storage will also be bad.)
- **Corky wine:** The most common flaw, *corkiness* comes across as a smell of damp cardboard that gets worse with air, and a diminished flavour intensity. It's caused by a bad cork, and any wine in a bottle that's sealed with a cork is at risk of it. Bad bottle.

Where to Buy Good Wine

Once you find the perfect wine, you're probably wondering where you should buy it. Buying wine in a store to drink later at home is great – not the least of which is that stores usually have a much bigger selection of wines than restaurants do, and they charge less for them. You can examine the bottles carefully and compare the labels. And you can drink the wine at home from the glass — and at the temperature — of your choosing. On the other hand, that big selection of wines in the shop can be downright daunting.

Depending on where you live, you can buy wine at all sorts of stores: supermarkets, wine superstores, discount warehouses, or small speciality wine shops. Each type of store has its own advantages and disadvantages in terms of selection, price, or service.

Ten clues for identifying a store where you should not buy wine

1. The dust on the wine bottles is more than 5mm thick.
2. Many of the white wines are dark gold or light-brown in colour.
3. The most recent vintage in the store is 1997.
4. The colours on all the wine labels have faded from bright sunlight.
5. It's warmer than a sauna inside.
6. Most of the bottles are standing up.
7. All the bottles are standing up!
8. The selection consists mainly of jug wines or cheap "Bag-in-the-Box" wines.
9. The June Wine of the Month is a three-year-old Beaujolais Nouveau.
10. The owner resembles the stern teacher who always hated you.

Supermarkets, superstores, and so on

In truly *open* wine markets, you can buy wine in supermarkets, like any other food product. Supermarkets and their large-scale brethren, discount warehouses, make wine accessible to everyone.

On the plus side, when wine is sold in supermarkets or discount stores, the mystique surrounding the product evaporates. And the prices, especially in large stores, are usually quite reasonable.

The downside of buying wine in these stores is that your selection is often limited to wines produced by large wineries that generate enough volume to sell to supermarket chains. And you'll seldom get any advice on which wines to buy. Basically, you're on your own.

TIP

The bottom line is that supermarkets and discount warehouses can be great places to buy everyday wine for casual enjoyment. But if what you really want is to learn about wine as you buy it, or if you want an unusually interesting variety of wines to satisfy your rapacious curiosity, you will probably find yourself shopping elsewhere.

TIP

Discount stores are good places to find *private label* wines — wines that are created especially for the chain, and that carry a brand name that's owned by the store. These wines usually are decent (but not great), and if you like the wines, they can be excellent value. Some of the "club" chains may also offer — in smaller quantities — higher-end wines than supermarkets do.

WARNING!

To guide you on your wine-buying journey, many shops offer plenty of *shelf-talkers* (small signs on the shelves that describe individual wines). Take these shelf-talkers with a very large grain of salt. They're often provided by the company selling the wine, which is more interested in convincing you to grab a bottle than in offering information to help you understand the wine. Most likely, you'll find flowery phrases, hyperbolic adjectives, impressive scores and safe, common-denominator stuff like "delicious with fish." The information will be biased and of limited value. Instead, find a knowledgeable person from the shop to help you, if at all possible, rather than rely on shelf-talkers.

Wine speciality shops

Wine speciality shops are small- to medium-sized shops that sell wine and other alcoholic drinks and, sometimes, wine books, corkscrews, wine glasses, and maybe a few speciality foods. The foods sold in wine shops tend to be gourmet items rather than just run-of-the-mill snack foods.

If you decide to pursue wine as a serious hobby, these shops are the places where you'll probably end up buying your wine because they offer many advantages that larger operations can't. For one thing, wine speciality shops almost always have wine-knowledgeable staffers on the premises. Also, you can usually find an interesting, varied selection of wines at all price levels.

Wine shops often organize their wines by country of origin and – in the case of classic wine countries, such as France, by region. Red wines and white wines are often in separate sections within these country areas. There may be a special section for Champagnes and other sparkling wines and another section for dessert wines. Some stores are now organizing their wine sections by style, such as "Aromatic Whites," "Powerful Reds," and so on. A few organize the wines according to grape varieties.

WARNING!

Over in a corner somewhere, often right by the door to accommodate quick purchases, there's usually a *cold box,* a refrigerated cabinet with glass doors where bottles of best-selling white and sparkling wines sit. Unless you really *must* have an ice-cold bottle of wine immediately, avoid the cold box. The wines in there are usually too cold and therefore may not be in good condition. You never know how long the bottle you select has been sitting there under frigid conditions, numbed lifeless.

Some wine shops have a special area (or even a super-special, temperature-controlled room) for the finer or more expensive wines. In some shops, it's a locked vault-like room. In others, it's the whole back area of the shop.

Near the front of the shop you may also see boxes or bins of special *sale* wines. Sometimes, sale wines are those the merchant is trying to unload because he's had them for too long, or they're wines that he got a special deal on (because the distributor is trying to unload them). When in doubt, try one bottle first before committing to a larger quantity.

Sale displays are usually topped with *case cards* – large cardboard signs that stand above the open boxes of wine – or similar descriptive material. They're not much more credible than shelf-talkers, but because case cards are a lot bigger, there's more of a chance that some useful information may appear on them.

Five questions you should ask in a wine shop

- If a wine costs more than about £7: What kind of storage has this wine experienced? Hemming and hawing on the part of the wine merchant should be taken to mean, "Poor."
- How long has this wine been in your shop? This question is especially important if the shop doesn't have a climate-control system.
- What are some particularly good buys this month? (Provided you trust the wine merchant, and you don't think he's dumping some overstocked, closeout wine on you.)
- If applicable: Why is this wine selling at such a low price? The merchant might know that the wine is too old or is otherwise defective; unless he comes up with a believable explanation, assume that's the case.
- Will this wine go well with the food I'm planning to serve? The more information about the recipe or main flavours you can provide, the better your chance of getting a good match.

Marrying Wine with Food

IN THIS ARTICLE

- *Predicting reactions between wines and foods*
- *Trying classic combos that still work*
- *Guiding principles for matchmakers*
- *Choosing wine when dining out*

Every now and then, a wine stops you dead in your tracks. It's so sensational that you lose all interest in anything but that wine. You drink it with intent appreciation, trying to memorize the taste. You wouldn't dream of diluting its perfection with a mouthful of food.

But 999 times out of 1,000, you drink your wine with food. Wine is meant to go with food. And good food is meant to go with wine.

Thousands of wines are in the world, and every one is different. And thousands of basic foods are in the world, each different – not to mention the infinite combinations of foods in prepared dishes (what you really eat). In reality, food-with-wine is about as simple an issue as boy-meets-girl.

The Dynamics of Food and Wine

Every dish is dynamic – it's made up of several ingredients and flavours that interact to create a (more or less) delicious whole. Every wine is dynamic in exactly the same way. When food and wine combine in your mouth, the dynamics of each change; the result is completely individual to each dish-and-wine combination.

When wine meets food, several things can happen:

- The food can exaggerate a characteristic of the wine. For example, if you eat walnuts (which are tannic) with a tannic red wine, such as a Bordeaux, the wine tastes so dry and astringent that most people would consider it undrinkable.
- The food can diminish a characteristic of the wine. Protein diminishes the impression of tannin, for example, and an overly-tannic red wine – unpleasant on its own – could be delightful with rare steak or roast beef.
- The flavour intensity of the food can obliterate the wine's flavour or vice versa. If you've ever drunk a big, rich red wine with a delicate fillet of sole, you've had this experience firsthand.
- The wine can contribute new flavours to the dish. For example, a red Zinfandel that's gushing with berry fruit can bring its berry flavours to the dish, as if another ingredient had been added.
- The combination of wine and food can create an unwelcome third-party flavour that wasn't in either the wine or the food originally; you get a metallic flavour when you eat plain white-meat turkey with red Bordeaux.
- The food and wine can interact perfectly, creating a sensational taste experience that is greater than the food or the wine alone. (This scenario is what you hope will happen every time you eat and drink, but it's as rare as a show-stopping dish.)

Fortunately, what happens between food and wine isn't haphazard. Certain elements of food react in predictable ways with certain elements of wine, giving you a fighting chance at making successful matches. The major components of wine (alcohol, sweetness, acid, and tannin) relate to the basic tastes of food (sweetness, sourness, bitterness, and saltiness) the same way that the principle of balance in wine operates: Some of the elements exaggerate each other, and some of them compensate for each other.

REMEMBER

Each wine and each dish has more than one component, and the simple relationships described can be complicated by other elements in the wine or the food. Whether a wine is considered tannic, sweet, acidic, or high in alcohol depends on its dominant component.

Tannic wines

Tannic wines include most wines based on the Cabernet Sauvignon grape (including red Bordeaux), northern Rhône reds, Barolo and Barbaresco, and any wine – white or red – that has become tannic from aging in new oak barrels. These wines can

- Diminish the perception of sweetness in a food
- Taste softer and less tannic when served with protein-rich, fatty foods, such as steak or cheese
- Taste less bitter when paired with salty foods
- Taste astringent, or mouth-drying, when drunk with spicy-hot foods

Sweet wines

Some wines that often have some sweetness include most inexpensive California white wines, White Zinfandel, many Rieslings (unless they're labeled dry or trocken), and medium-dry Vouvray. Sweet wines also include dessert wines such as Port, sweetened Sherries, and late-harvest wines. These wines can

- Taste less sweet, but fruitier, when matched with salty foods
- Make salty foods more appealing
- Go well with sweet foods

Acidic wines

Acidic wines include most Italian white wines; Sancerre, Pouilly-Fumé, and Chablis; traditionally-made red wines from Rioja; most dry Rieslings; and wines based on Sauvignon Blanc that are fully dry. These wines can

- Taste less acidic when served with salty foods
- Taste less acidic when served with slightly sweet foods
- Make foods taste slightly saltier
- Counterbalance oily or fatty heaviness in food

High-alcohol wines

High alcohol wines include many California wines, both white and red; southern Rhône whites and reds; Barolo and Barbaresco; fortified wines such as Port and Sherry; and most wines produced from grapes grown in warm climates. These wines can

- Overwhelm lightly flavoured or delicate dishes
- Go well with slightly sweet foods

The fifth wheel

Common wisdom was that humans can perceive four basic tastes: sweet, sour, salty, and bitter. But people who study food have concluded that a fifth taste exists, and there may be many more than that. The fifth taste is called umami (pronounced oo MAH me), and it's associated with a savoury character in foods. Shellfish, oily fish, meats, and cheeses are some foods high in umami taste.

Umami-rich foods can increase the sensation of bitterness in wines served with them. To counteract this effect, try adding something salty (such as salt itself) or sour (such as vinegar) to your dish. Although this suggestion defies the adage that vinegar and wine don't get along, the results are the proof of the pudding.

Birds of a Feather, or Opposites Attract?

Two principles can help in matching wine with food: the complementary principle and the contrast principle. The complementary principle involves choosing a wine that is similar in some way to the dish you plan to serve, while the contrast principle (not surprisingly) involves combining foods with wines that are dissimilar to them in some way.

TIP

The characteristics of a wine that can either resemble or contrast with the characteristics of a dish are

- **The wine's flavours:** Earthy, herbal, fruity, vegetal, and so on
- **The intensity of flavour in the wine:** Weak flavour intensity, moderately flavourful, or very flavourful
- **The wine's texture:** Crisp and firm, or soft and supple
- **The weight of the wine:** Light-bodied, medium-bodied, or full-bodied

TIP

In order to apply either the complementary or contrast principle, you have to have a good idea of what the food is going to taste like and what various wines taste like. That second part can be a real stumbling block for people who don't devote every ounce of their free energy to learning about wine. The solution is to ask your wine merchant. A retailer may not have the world's greatest knack in wine and food pairings (then again, he or she might), but at least he should know what his wines taste like.

The complementary principle

You probably use the complementary principle often without realizing it: You choose a light-bodied wine to go with a light dish, a medium-bodied wine to go with a fuller dish, and a full-bodied wine to go with a heavy dish. Some other examples of the complementary principle in action are

- **Dishes with flavours that resemble those in the wine.** Think about the flavours in a dish the same way you think about the flavours in wine – as families of flavours. If a dish has mushrooms, it has an earthy flavour; if it has citrus or other elements of fruit, it has a fruity flavour (and so on). Then consider which wines would offer their own earthy flavour, fruity flavour, herbal flavour, spicy flavour, or whatever. The earthy flavours of white Burgundy complement risotto with mushrooms, for example, and an herbal Sancerre complements chicken breast with fresh herbs.

- **Foods with texture that's similar to that of the wine.** A California Chardonnay with a creamy, rich texture could match the rich, soft texture of lobster, for example.
- **Foods and wines whose intensity of flavour match.** A very flavourful Asian stir-fry or Tex-Mex dish would be at home with a very flavourful, rather than subtle, wine.

The contrast principle

The contrast principle seeks to find flavours or texture in a wine that aren't in a dish but that would enhance it. A dish of fish or chicken in a rich cream and butter sauce, for example, may be matched with a dry Vouvray, a white wine whose crispness (thanks to its uplifting, high acidity) would counterbalance the heaviness of the dish. A dish with earthy flavours such as portobello mushrooms and fresh fava beans (or potatoes and black truffles) may contrast nicely with the pure fruit flavour of an Alsace Riesling.

You also apply the contrast principle every time you decide to serve simple food, like unadorned lamb chops or hard cheese and bread, with a gloriously complex aged wine.

White wine with fish, red with meat – or not?

As guidelines go, pairing white wine with fish and red wine with meat isn't a bad one. But it's *guideline*, not rule. Anyone who slavishly adheres to this generalization deserves the boredom of eating and drinking exactly the same thing every day of his life! Do you want a glass of white wine with your burger? Go ahead, order it. You're the one who's doing the eating and drinking, not your friend and not the server who's taking your order.

Even if you're a perfectionist who's always looking for the ideal food and wine combination, you'll find yourself wandering from the guideline. The best wine for a grilled salmon steak is probably red – like a Pinot Noir or a Bardolino – and not white at all. Veal and pork do equally well with red or white wines, depending on how the dish is prepared. And what can be better with hot dogs on the grill than a cold glass of rosé?

No one is going to arrest you if you have white wine with everything, or red wine with everything, or even Champagne with everything! There are no rules.

The Wisdom of the Ages: Classic Pairings

No matter how much you value imagination and creativity, there's no sense reinventing the wheel. In wine-and-food terms, it pays to know the classic pairings because they work, and they're a sure thing.

Here are some famous and reliable combinations:

- Oysters and traditional, unoaked Chablis
- Lamb and red Bordeaux (try Chianti with lamb, too)
- Port with walnuts and Stilton cheese
- Salmon with Pinot Noir
- Amarone with Gorgonzola cheese
- Grilled fish with Vinho Verde
- Foie gras with Sauternes or with late-harvest Gewürztraminer
- Braised beef with Barolo
- Dry amontillado Sherry with soup
- Grilled chicken with Beaujolais
- Toasted almonds or green olives with fino or manzanilla Sherry
- Goat cheese with Sancerre or Pouilly-Fumé
- Dark chocolate with California Cabernet Sauvignon

Reliable Wine Choices When Ordering in A Restaurant

Not sure what wine to order at a restaurant? Keep these following pointers in mind:

- To accompany delicately flavoured fish or seafood, you want a crisp, dry white wine that isn't very flavourful. Order Soave, Pinot Grigio, or Sancerre.
- When eating mussels and other shellfish, you want a dry white wine with assertive flavour. Order Sauvignon Blanc from South Africa or New Zealand.
- For simple poultry, risotto, and dishes that are medium in weight, you want a medium-bodied, characterful, dry white wine. Order Mâcon-Villages, St.-Véran, or Pouilly-Fuissé.
- For lobster or rich chicken entrées, you want a full-bodied, rich white wine. Order Californian or Australian Chardonnay.
- To accompany meaty fish, veal, or pork entrées, you want a full-bodied white wine with a honeyed, nutty character. Order Meursault.
- For Asian-inspired dishes, you want a medium-dry white wine. Order Chenin Blanc, Vouvray, or German Riesling.
- An easy-drinking, inexpensive red is perfect with roast chicken. Order Beaujolais (especially from a reputable producer, like Louis Jadot, Joseph Drouhin, or Georges Duboeuf).
- When you want a versatile, flavourful, relatively inexpensive red that can stand up to spicy food, order California red Zinfandel.
- A lighter red that's delicious and young and works with all sorts of light- and medium-intensity foods. Order Oregon or California Pinot Noir.
- With simple cuts of steak, you want the basic French version of Pinot Noir. Order Bourgogne Rouge.
- When you want a dry, spicy, grapey, and relatively inexpensive red wine that's perfect with pizza, order Barbera or Dolcetto.
- When you want a very dry, medium-bodied red that's great with lots of foods, order Chianti Classico.

Buying and serving bubbly

Sparkling wine is best cold, at about F 7° to 8°C (45°), although some people prefer it less cold (11°C). The colder temperature helps the wine hold its effervescence – and the wine warms up so quickly in the glass, anyway. Because older Champagnes and Vintage Champagnes are more complex, you can chill them less than young, nonvintage Champagne or sparkling wine.

Never leave an open bottle of sparkling wine on the table unless it's in an ice bucket (half cold water, half ice) because it will warm up quickly. Use a sparkling wine stopper to keep leftover bubbly fresh for a couple of days—in the fridge, of course.

If you're entertaining, you should know that the ideal bottle size for Champagne is the magnum, which is equivalent to two bottles. The larger bottle enables the wine to age more gently in the winery's cellar. Magnums (or sometimes double magnums) are usually the largest bottles in which Champagne is fermented; all really large bottles have had finished Champagne poured into them, and the wine is therefore not as fresh as it is in a magnum or a regular bottle.

Be wary of half-bottles (375 ml) and – chancier yet – splits (187 ml)! Champagne in these small bottles is often not fresh. If you're given a small bottle of Champagne or any sparkling wine as a wedding present, for example, open it at the first excuse; don't keep it around for a year waiting for the right occasion!

Champagne and other good, dry sparkling wines are extremely versatile with food – and they're the essential wine for certain kinds of foods. For example, no wine goes better with egg dishes than Champagne. Indulge yourself next time that you have brunch. And when you're having spicy Asian cuisine, try sparkling wine. No wine matches up better with spicy Chinese or Indian food!

Fish, seafood, pasta (but not with tomato sauce), risotto, and poultry are excellent with Champagne and sparkling wine. If you're having lamb (pink, not well-done) or ham, pair rosé Champagne with it. With aged Champagne, chunks of aged Asiago, aged Gouda, or Parmesan cheese go extremely well.

Don't serve a dry brut (or extra dry) sparkling wine with dessert. These styles are just too dry. With fresh fruit and desserts that aren't too sweet, try a demi-sec Champagne. With sweeter desserts (or wedding cake!), try Asti.

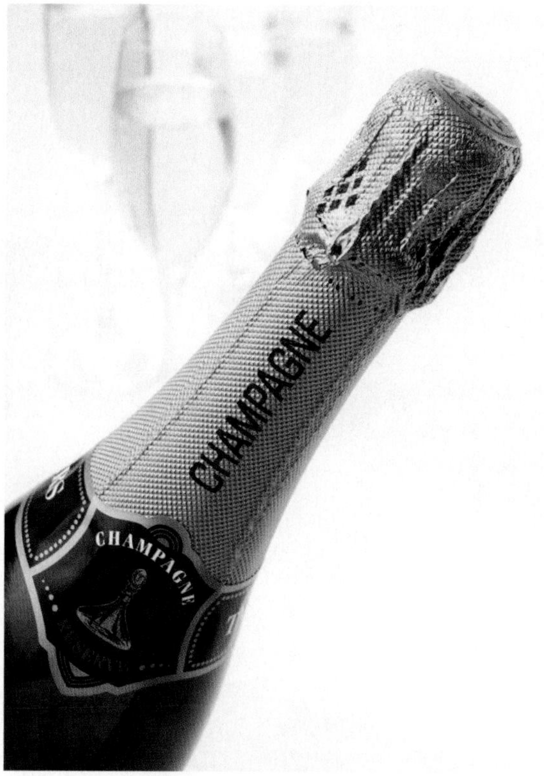

Restaurant Wine Tips

Drinking wine in a restaurant requires so many decisions that you really do need a guidebook. Should you leave the wine in an ice bucket? What should you do if the wine is bad? And can you bring your own wine? Let the tips in this feature guide you:

Can I kick the ice-bucket habit?

Most servers assume that an ice bucket is necessary to chill white wines and sparkling wines. But sometimes the bottle is already so cold when it comes to you that the wine would be better off warming up a bit on the table. If your white wine goes into an ice bucket and you think it's getting ***too*** cold, remove it from the bucket, or have the waiter remove it. Just because that ice bucket is sitting there on your table (or next to your table) doesn't mean that your bottle has to be in it!

What's with these tiny glasses?

When various glasses are available, you can exercise your right to choose a different glass from the one you were given. If the restaurant's red wine glass is quite small, a stemmed water glass might be more appropriate for the red wine.

Should the wine "breathe"?

If a red wine you ordered needs aeration to soften its harsh tannins, merely pulling the cork will be practically useless in accomplishing that (because the air space at the neck of the bottle is too small). Decanting the bottle or pouring the wine into glasses is the best tactic.

TIP

Sometimes, a red wine that's a bit too warm can benefit from five or ten minutes in an ice bucket. (But be careful! It can get too cold very quickly.) If the server acts as if you're nuts to chill a red wine, ignore him.

Where's my bottle?

Have your bottle of wine on or near your table, not out of your reach. That way, you can look at the label, and you don't have to wait for the server to remember to refill your glasses, either.

What if the bottle is bad?

Refuse any bottle that tastes or smells unpleasant (unless you brought it yourself!). A good restaurateur will always replace the wine, even if he thinks there's nothing wrong with it.

May I bring my own wine?

Some restaurants allow you to bring your own wine – especially if you express the desire to bring a special wine, or an older wine. Restaurants will usually charge a ***corkage*** fee (a fee for wine service, use of the glasses, and so on) that can vary from £6 to £15 a bottle, depending on the attitude of the restaurant. Call ahead to determine whether it's possible to bring wine (in some places, the restaurant's license prohibits it) and to ask what the corkage fee is.

What if I'm travelling abroad?

If you journey to countries where wine is made, such as France, Italy, Germany, Switzerland, Austria, Greece, Spain, or Portugal, by all means try the local wines. They will be fresher than the imports, in good condition, and the best values on the wine list. It doesn't make sense to order French wines, such as Bordeaux or Burgundy, in Italy, for example, or California Cabernets in Paris.

WARNING!

You should never bring a wine that is already on the restaurant's wine list; it's cheap and insulting. (Call and ask the restaurant when you're not sure whether the wine is on its list.)

French Wines and the Legendary Bordeaux

IN THIS ARTICLE

- *Exploring France's fine wine regions*
- *Getting the inside scoop on Bordeaux wines*

When most people think of France, they think of wine. Why has France become the most famous place in the world for wine? For one thing, the French have been making wine – and doing it right – for a long time. Equally important is French *terroir,* the magical combination of climate and soil that, when it clicks, yields grapes that make breathtaking wines.

France is the model, the standard setter, for all the world's wines. If imitation is the sincerest form of flattery, French winemakers have had good reason to blush for a long time.

To really know wine, you must know French wine – French wines are that important in the wine world. Likewise, you must know Bordeaux to know French wine.

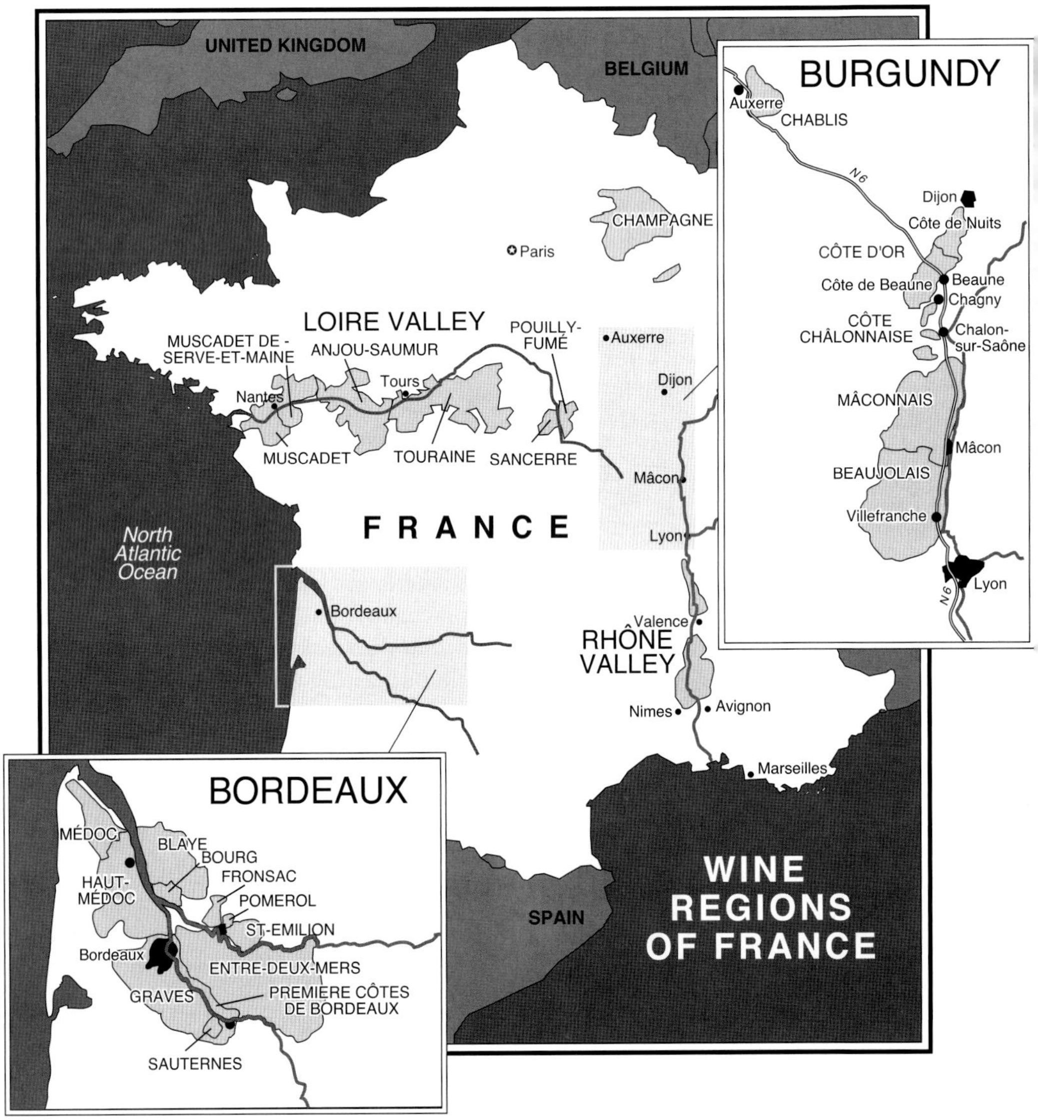

The wine regions of France.

France's Wine Regions

France has five wine regions (see thr map) that are extremely important for the quality and renown of the wines they produce, and it has several other regions that make interesting wines worth knowing about. Each region specializes in certain grape varieties for its wines, based on climate, soil, and local tradition.

The three major regions for red wine are Bordeaux, Burgundy, and the Rhône; for white wines, Burgundy is again a major region, along with the Loire and Alsace.

Introducing the Bordeaux Wine Region

Bordeaux is a wine region in western France named after the fourth-largest French city. Because the Bordeaux region is situated on the Atlantic coast, it has a maritime climate, with warm summers and fairly mild winters. The maritime weather brings rain, often during harvest time. The weather varies from year to year, and the character and quality of the vintages therefore also vary; when all goes well, such as in 1996, 2000, and 2005, the wines can be great.

Bordeaux's reputation as one of the greatest wine regions in the world revolves around the legendary red wines of Bordeaux – *grands vins* (great wines) made by historic *châteaux* (wine estates) and capable of improving for many decades. Prices for these wines run to well over £315 a bottle for Château Pétrus — one of the most expensive red Bordeaux wines.

These legendary wines represent the pinnacle of a red Bordeaux pyramid; quantitatively, they're only a very small part of the region's red wine production, which also includes medium-priced and even inexpensive wines. Middle-level Bordeaux reds are ready to drink within 10 to 15 years of the harvest.

Most fine Bordeaux wines, both red and white, start at about £15 a bottle when they're first available, two or three years after the vintage. The least expensive Bordeaux reds, which can cost as little as £5 a bottle, are enjoyable young, within two to five years of the vintage date.

The Subregions of Red Bordeaux

Two distinct red wine production zones exist within the Bordeaux region; these two areas have come to be called the Left Bank and the Right Bank – just as in Paris. While many of the least expensive Bordeaux reds are blended from grapes grown all through the Bordeaux region – and thus carry the region-wide AOC designation, Bordeaux – the better wines come from specific AOC *districts* or AOC *communes* that are located in either the Right Bank or the Left Bank.

The Left Bank vineyards lie west of the Garonne River and the Gironde Estuary into which the Garonne empties. The Right Bank vineyards lie east and north of the Dordogne River (the more northerly of the two rivers depicted in the inset), and east of the Gironde Estuary. (The middle ground in between the two rivers is more important for white wine than for red.)

Of the various wine districts on the Left Bank and the Right Bank, four (two on each bank) are the most important:

- Left Bank (the western area): Haut-Médoc, Pessac-Léognan
- Right Bank (the eastern area): St-Emilion, Pomerol

The Left Bank and the Right Bank differ mainly in soil composition: Gravel predominates on the Left Bank, and clay prevails on the Right Bank. As a result, Cabernet Sauvignon, which has an affinity for gravel, is the principal grape variety in the Haut-Médoc *(oh meh doc)* and Pessac-Léognan *(pay sac lay oh nyahn)*. Merlot, which does well in clay, dominates the St-Emilion *(sant em eel yon)* and Pomerol *(pohm eh roll)* wines. (Both areas grow Cabernet Sauvignon *and* Merlot, as well as Cabernet Franc and two less significant grapes.)

Left Bank and Right Bank Bordeaux wines are markedly different from one another. But wines from the Haut-Médoc and Pessac-Léognan are quite similar. Likewise, it can be difficult to tell the difference among wines from Pomerol and St-Emilion (on the Right Bank).

Each bank – in fact, each of the four districts – has its avid fans. The more established Left Bank generally produces austere, tannic wines with more pronounced blackcurrant flavour. Left Bank wines usually need many years to develop and will age for a long time, often for decades – typical of a Cabernet Sauvignon-based wine.

TIP

Bordeaux wines from the Right Bank are better introductory wines for the novice Bordeaux drinker. Because they're mainly Merlot, they're more approachable; you can enjoy them long before their Left Bank cousins, often as soon as five to eight years after the vintage. They're less tannic, richer in texture, and plummier in flavour, and they generally contain a bit more alcohol than Left Bank reds.

The Médoc Mosaic

Historically, the Haut-Médoc has always been Bordeaux's most important district, and it deserves special attention. The Haut-Médoc is actually part of the Médoc peninsula. Médoc is frequently used as an umbrella term for the combined districts of Médoc and Haut-Médoc (the two districts that occupy the Médoc peninsula).

Of the two districts, the Haut-Médoc, in the south, is by far the more important for wine. The Haut-Médoc itself encompasses four famous wine communes: St. Estèphe *(sant eh steff)*, Pauillac *(poy yac)*, St-Julien *(san jhoo lee ehn)*, and Margaux *(mahr go)*.

The following list gives a general description of each commune's wines.

- **St-Estèphe:** Firm, tannic, earthy, chunky, and slow to mature; typical wine – Château Montrose
- **Pauillac:** Rich, powerful, firm, and tannic, with black-currant and cedar aromas and flavours; very long-lived; home of three of Bordeaux's most famous wines – Lafite-Rothschild, Mouton-Rothschild, and Latour
- **St-Julien:** Rich, flavourful, elegant and finesseful, with cedary bouquet; typical wine – Château Ducru-Beau-caillou
- **Margaux:** Fragrant, supple, harmonious, with complex aromas and flavours; typical wine – Château Palmer

Two other communes in the Haut-Médoc – Listrac *(lee strahk)* and Moulis *(moo lees)* – make less well-known wines. Vineyards in the Haut-Médoc that aren't located in the vicinity of these six communes carry the district-wide appellation, *Haut-Médoc,* rather than that of a specific commune.

REMEMBER

The names of these districts and communes are part of the official name of wines made there, and appear on the label.

Classified information

Have you ever wondered what a wine expert was talking about when he smugly pronounced a particular Bordeaux second growth? Wonder no more. He's talking about a château (as wine estates are called in Bordeaux) that made the grade about 150 years ago.

Back in 1855, when an Exposition (akin to a World's Fair) took place in Paris, the organizers asked the Bordeaux Chamber of Commerce to develop a classification of Bordeaux wines. The Chamber of Commerce delegated the task to the Bordeaux wine brokers, the people who buy and re-sell the wines of Bordeaux. These merchants named 61 top red wines – 60 from the Médoc and one from what was then called Graves (and today is known as Péssac-Leognan). According to the prices fetched by the wines at the time and the existing reputations of the wines, they divided these 61 wines into five categories, known as crus or growths. (In Bordeaux, a cru refers to a wine estate.) Their listing is known as the classification of 1855; to this day, these classified growths enjoy special prestige among wine lovers. (The Bordeaux wine brokers also classified Sauternes, the great Bordelais dessert wine).

Bordeaux to Try

TIP

If you're curious to try a prestigious red Bordeaux, let this list guide you. In addition to all five first growths listed in the previous section, try the following classified growths from the Médoc, as well as some wines from the three other principal districts: Pessac-Léognan, St-Emilion, and Pomerol.

Médoc wines:

Château Léoville-Las-Cases
Château Léoville-Barton
Château Rauzan-Ségla
Château Palmer
Château Cos D'Estournel
Château Léoville-Poyferré
Château Pontet-Canet
Château Haut-Batailley
Château Duhart-Milon- Rothschild
Château Clerc-Milon
Château Gruaud-Larose
Château Pichon-Lalande
Château Lagrange
Château Pichon-Baron
Château d'Armailhac
Château Prieuré-Lichine
Château Malescot-St-Exupéry
Château Calon-Ségur
Château Lynch-Bages
Château Montrose
Château Ducru-Beaucaillou
Château Grand-Puy-Lacoste
Château La Lagune
Château Branaire-Ducru
Château Batailley
Château Talbot

Pessac-Léognan wines:

Château La Mission-Haut-Brion
Château Pape-Clément
Château La Tour-Haut-Brion
Château Haut-Bailly
Domaine de Chevalier
Château Smith-Haut-Lafitte
Château de Fieuzal
Château La Louvière

Pomerol wines:

Château Pétrus*
Château Lafleur*
Château Latour à Pomerol
Château Certan de May
Château Lafleur-Pétrus
Château Trotanoy
Château Clinet
Vieux-Château-Certan
Château Gazin
Château L'Evangile
Château La Fleur de Gay
Château La Conseillante
Château l'Eglise Clinet

** Very expensive*

St-Emilion wines:

Château Cheval Blanc
Château Ausone
Château Figeac
Château Pavie- Macquin
Château La Dominique
Château Grand Mayne
Château Troplong Mondot
Château Magdelaine
Château Canon-La-Gaffelière
Château L'Arrosée
Château Clos Fourtet

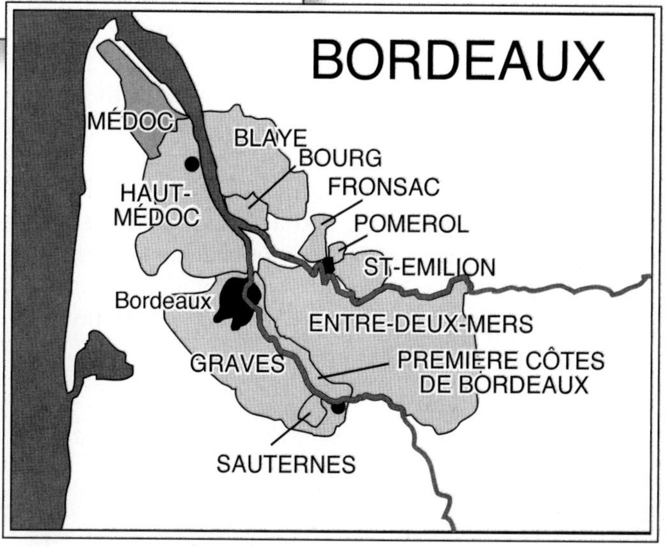

The Value End of the Bordeaux Spectrum

Cru Bourgeois: The middle class of the Médoc

The wines in the *Cru Bourgeois* category are considered just a bit less noble than the *Grands Crus Classé* wines, but sell at considerably lower prices, mainly in the £11 to £22 range. Some of them are even as good as the lesser quality classified growths.

In 2005, the Bordeaux Chamber of Commerce reclassified the *Cru Bourgeois* wines. Nine chateaux are now in the highest level, *Cru Bourgeois Exceptionnel*:

Château Chasse-Spleen
Château Poujeaux
Château Haut-Marbuzet
Château de Pez
Château Siran
Château Phélan-Ségur
Château Les Ormes-de-Pez
Château Labegorce-Zédé
Château Potensac

Château Gloria, from St.-Julien, a wine often compared to *Grands Crus Classé* wines in quality, refused to apply for membership as a *Cru Bourgeois* wine, but critics generally agree that it would have been awarded *Cru Bourgeois Exceptionnel* status had it done so. In addition to the preceding wines, the following Cru Bourgeois wines are recommended:

Château Monbrison
Château d'Angludet
Château Coufran
Château Haut-Beauséjour
Château Fourcas-Hosten
Château Bel Air
Château Meyney
Château Sociando Mallet
Château Lanessan
Château Loudenne
Château Monbousquet
Château Greysac

Fronsac and Canon-Fronsac

When you leave the Médoc peninsula and the city of Bordeaux and cross over the Dordogne River into the Right Bank region, the first wine districts you encounter on your left are Fronsac and Canon-Fronsac. Of the lesser Bordeaux appellations, Canon-Fronsac and Fronsac have the highest reputations for quality, and they're priced accordingly, in the £12 to £18 range. Like St.-Emilion and Pomerol, Fronsac and Canon-Fronsac produce only red wines, and Merlot is the dominant grape variety.

TIP

As you may have suspected, the best buys in Bordeaux wines aren't the illustrious classified growths. For really good values (and wines that you can drink within a few years of the vintage), look for Bordeaux wines not included in the 1855 classification.

Petits Chateaux

Petits chateaux is the general, catchall term for the huge category of reasonably priced wines throughout the entire Bordeaux region that have never been classified. The term is somewhat of a misnomer because it suggests that the wines come from a specific chateau or vineyard estate; in fact, many *petits chateaux* do come from specific estates, but not all.

Some of these wines use grapes that have been sourced from all over the region, and others come from specific appellations. Ten petit chateau red appellations, all on the Right Bank, are noteworthy:

Côtes de Bourg	Puisseguin-St.-Emilion
Premières Côtes de Blaye	Lussac-St.-Emilion
Côtes de Castillon	Montagne-St.-Emilion
Côtes de Francs	St.-Georges-St.-Emilion
Lalande de Pomerol	Premières Côtes de Bordeaux

The wines from the Côtes de Bourg, Premières Côtes de Blaye, and Lalande de Pomerol are particularly good value. The wines from all ten appellations are mainly in the £7 to £13 range.

All these wines are primarily Merlot. They're generally fruitier, have softer tannins, and are ready to drink sooner than the Cabernet Sauvignon-dominated wines of the Left Bank. *Petits châteaux* are the Bordeaux wines of choice when you're looking for a young, inexpensive, approachable Bordeaux with dinner.

Generic Bordeaux

Red Bordeaux wines with no specific appellation carry the general "Bordeaux" or "Bordeaux Supérieur" appellations. Their grapes are predominantly Merlot and can grow anywhere throughout the Bordeaux region. These are fairly light-bodied wines ("mild-mannered reds") that sell for £5 to £7. Sometimes the labels identify the wines as specifically Merlot or Cabernet Sauvignon. Two leading brands are Mouton-Cadet and Michel Lynch. Generic Bordeaux from good vintages, such as 2000 and 2005, can be really excellent buys.

Practical Advice on Drinking Red Bordeaux

Because the finest red Bordeaux wines take many years to develop, they're often not good choices in restaurants, where the vintages available tend to be fairly recent. And when mature Bordeaux wines are available in restaurants, they're usually extremely expensive. Order a lesser Bordeaux when you're dining out and save the best ones for drinking at home.

TIP

Fine recent Bordeaux vintages are 2005 (which shows promise of being truly great), 2000, 1996, 1995, 1990, 1989, 1986, and 1982.

TIP

Red Bordeaux wines go well with lamb, venison, simple roasts, and hard cheeses, such as Comte, Gruyère, or Cheddar. If you plan to serve a fine red Bordeaux from a good but recent vintage, you should decant it at least an hour before dinner and let it aerate; serve it at about 17° to 19°C. Better yet, if you have good storage conditions, save your young Bordeaux for a few years – it will only get better.

Bordeaux Also Comes in White

White Bordeaux wine comes in two styles, dry and sweet. The dry wines themselves are really in two different categories: inexpensive wines for enjoying young, and wines so distinguished and age-worthy that they rank among the great dry white wines of the world.

Two areas of the Bordeaux region are important for white wine production:

- The large district south of the city of Bordeaux is known as the Graves (grahv). The Graves district and the Pessac-Léognan district, directly north (around the city of Bordeaux) are home to the finest white wines of Bordeaux, both dry and sweet.
- In the middle ground between the Garonne and Dordogne Rivers, east of Graves and Pessac-Léognan, a district called Entre-Deux-Mers (ahn treh-douh-mare) is also known for its dry, semi-dry, and sweet white Bordeaux wines.

A few white wines also come from the predominantly red-wine Haut-Médoc district, such as the superb Pavillon Blanc du Château Margaux. Although special and expensive, they qualify only for simple Bordeaux blanc appellation.

Sauvignon Blanc and Sémillon, in various combinations, are the two main grape varieties for the top white Bordeaux. It's a fortunate blend: The Sauvignon Blanc component offers immediate charm in the wine, while the slower-developing Sémillon gives the wine a viscous quality and depth, enabling it to age well. In general, a high percentage of Sémillon in the wine is a good indicator of the wine's ageworthiness. Many inexpensive white Bordeaux — and a few of the best wines — are entirely Sauvignon Blanc.

The top dry white Bordeaux wines are crisp and lively when they're young, but they develop richness, complexity, and a honeyed bouquet with age. In good vintages, the best whites need at least ten years to develop and can live many years more.

The following list highlights the top 12 white wines of Pessac-Léognan and Graves, in rough order of preference, and shows their grape blends. The wines are separated into an A and B group because the four wines in the first group literally are in a class by themselves, quality-wise; they possess not only more depth and complexity but also more longevity than other white Bordeaux. Their prices reflect that fact – the A group wines range from £50 to £189 per bottle, whereas the B group wines cost between £18 and £44.

Group A:

- Château Haut-Brion Blanc: Sémillon, 50 to 55%; Sauvignon Blanc, 45 to 50%
- Château Laville-Haut-Brion: Sémillon, 60%; Sauvignon Blanc, 40%
- Domaine de Chevalier: Sauvignon Blanc, 70%; Sémillon, 30%
- Château Pape-Clément: Sémillon, 45%; Muscadelle, 10%; Sauvignon Blanc, 45%

Group B:

- Château de Fieuzal: Sauvignon Blanc, 50 to 60%; Sémillon, 40 to 50 %
- Château Smith-Haut-Lafitte: Sauvignon Blanc, 100%
- Clos Floridene: Sémillon, 70%; Sauvignon Blanc, 30%
- Château La Louvière: Sauvignon Blanc, 70%; Sémillon, 30%
- Château La Tour-Martillac: Sémillon, 60%; Sauvignon Blanc, 30%; other, 10%
- Château Couhins-Lurton: Sauvignon Blanc, 100%
- Château Malartic-Lagravière: Sauvignon Blanc, 100%
- Château Carbonnieux: Sauvignon Blanc, 65%; Sémillon, 35%

Burgundy: The Other Great French Wine

IN THIS ARTICLE

- *Exploring Burgundy's wine districts*
- *Looking at climate and soil*
- *Describing Burgundy's wines*

Burgundy, a wine region in eastern France, southeast of Paris, stands shoulder-to-shoulder with Bordeaux as one of France's two greatest regions for dry, nonsparkling wines.

Unlike Bordeaux, Burgundy's fame is split nearly equally between its white and red wines. Also unlike Bordeaux, good Burgundy is often scarce. The reason is simple: Not counting Beaujolais (which is technically Burgundy, but really a separate type of wine), Burgundy produces only 25 per cent as much wine as Bordeaux.

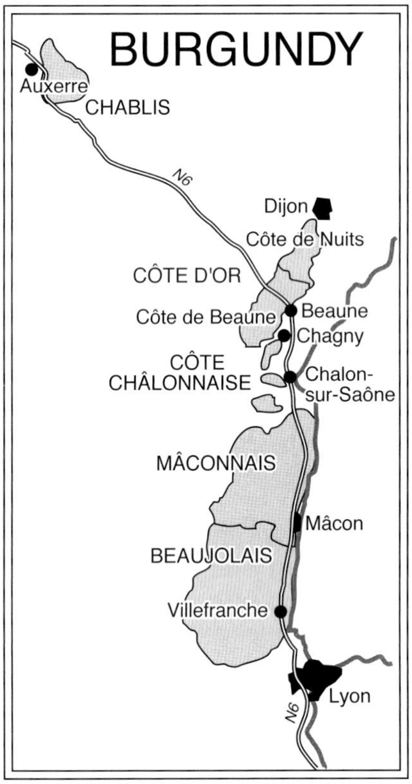

Districts, Districts Everywhere

Burgundy has five districts, all of which make quite distinct wines. The districts, from north to south, are

- Chablis *(shah blee)*
- The Côte d'Or *(coat dor)*
- The Côte Chalonnaise *(coat shal oh naze)*
- The Mâconnais *(mack coh nay)*
- Beaujolais *(boh jhoe lay)*

The heart of Burgundy, the *Côte d'Or* (which literally means *golden slope*), itself has two parts: Côte de Nuits *(coat deh nwee)* in the north and the Côte de Beaune *(coat deh bone)* in the south.

The Chablis district makes only white wines, and the Mâconnais makes mainly white wines. Beaujolais makes almost exclusively red wines; even though Beaujolais is part of Burgundy, Beaujolais is an entirely different wine, because it is made with the Gamay grape rather than Pinot Noir. The same is true of Mâcon Rouge, from the Mâconnais district; even the small amount that's made from Pinot Noir rather than from Gamay doesn't resemble more northerly red Burgundies. (Actually, very little red Mâcon is exported; the world sees mainly white Mâcon.)

REMEMBER

The term *red Burgundy* refers primarily to the red wines of the Côte d'Or and also to the less well-known – and less expensive – red wines of the Côte Chalonnaise. Likewise, when wine lovers talk about *white Burgundy,* they're usually referring just to the white wines of the Côte d'Or and the Côte Chalonnaise. They'll use the more precise names, Chablis and Mâcon, to refer to the white wines of those parts of Burgundy. On the other hand, when wine lovers talk about the region, Burgundy, they could very well be referring to the whole shebang, including Beaujolais, or all of Burgundy *except* Beaujolais. It's an imprecise language.

Burgundy versus Bordeaux

Burgundy's vineyards are more fragmented than Bordeaux's. The soils of the Burgundy region vary from hillside to hillside and even from the middle of each hill to the bottom. You can find two different vineyards growing the same grape but making distinctly different wines only two meters apart from each other across a dirt road!

Burgundy is also a region of much smaller vineyard holdings than Bordeaux. The few large vineyards that do exist in Burgundy have multiple owners, with some families owning only two or three rows of vines in a particular vineyard. The typical Burgundy winemaker's production varies from 50 cases to 1,000 cases of wine a year, per type – far from enough to satisfy wine lovers all over the world. Compare that to Bordeaux, where the average château owner makes 15,000 to 20,000 cases of his principal wine annually.

From the Regional to the Sublime: Climate and Soil

Burgundy has a *continental* climate (warm summers and cold winters) and is subject to localized summer hailstorms that can damage the grapes and cause rot. The soil is mainly limestone and clay. Burgundy's *terroir* is particularly suited to the two main grape varieties of the region, Pinot Noir (for red Burgundy) and Chardonnay (for white Burgundy). In fact, nowhere else in the entire world does the very fickle, difficult Pinot Noir grape perform better than in Burgundy.

In the southerly Beaujolais district of Burgundy, the soil becomes primarily granitic but also rich in clay and sand, very suitable for the Gamay grape of this area.

Because soils vary so much in Burgundy, a wine's specific vineyard site becomes extremely relevant to the taste, quality, and price of that wine. A wine made from a tiny vineyard with its own particular characteristics is unique, more precious and rare than a wine blended from several vineyards or a wine from a less favored site.

Site matters

The AOC structure for Burgundy wines recognizes the importance of site. While there are region-wide AOCs, district-wide AOCs, and commune AOCs – just as in Bordeaux – *there are also AOC names that refer to individual vineyards.* In fact, some of these vineyards are recognized as better than others: Some of them are *premier cru (prem yay crew)*, meaning first growth, while the very best are *grand cru,* meaning great growth.

The availability and price of each category's wines vary in the following ways:

- The two broadest categories – regional and district place-names – account for 65 per cent of all Burgundy wines. These wines retail for £6 to £15 a bottle. (You *can* buy affordable Burgundies at this level.)
- Commune-specific (also referred to as *village*) wines, such as Pommard or Gevrey-Chambertin, make up 23 per cent of Burgundy and are in the £15to £34 per bottle price range. Fifty-three communes in Burgundy have AOC status.
- Premier crus, such as Meursault Les Perrières or Nuits-St. Georges Les Vaucrains, account for 11 per cent of Burgundy wines; 561 vineyards have premier cru status. Most of these wines cost from £20 to £50 per bottle — but a few cost over £100 per bottle.
- The 31 grand crus, such as Chambertin, represent only *1 per cent* of Burgundy's wines. Prices for grand cru Burgundies – both red and white – start at around £50 and can go to well over £500 a bottle for Romanée-Conti, normally Burgundy's most expensive red wine.

TIP

Thankfully, you can usually tell the difference between a premier cru and a grand cru Burgundy by looking at the label. Premier cru wines tend to carry the name of their commune plus the vineyard name – most often in the same-sized lettering – on the label and, often, the words *Premier Cru* (or *1er Cru).* If a vineyard name is in smaller lettering than the commune name, the wine is generally not a premier cru but a wine from a single-vineyard site in that commune. (Not all single vineyards have premier cru status.) Grand cru Burgundies carry only the name of the vineyard on the label.

The taste of fine red Burgundy

Red Burgundy is paler than Bordeaux, with a less dense ruby or garnet colour, because Pinot Noir grapes don't have nearly as much pigmentation as Cabernet Sauvignon or Merlot grapes. It's rather full-bodied in terms of its alcohol and is relatively low in tannin. The characteristic aroma is of red or black berries and woodsy, damp-earth, or mushroomy scents. When a red Burgundy ages, it often develops a silky texture, richness, and a natural sweetness of fruit flavours; sometimes a bouquet of leather, coffee, and game emerges.

With some exceptions (for example, a powerful wine from a great vintage, such as 2002 or 1996), red Burgundy should be consumed within ten years of the vintage – and even sooner in a weaker vintage.

The Côte d'Or: The Heart of Burgundy

The Côte d'Or, a narrow 40-mile stretch of land with some of the most expensive real estate in the world, is the region where all the famous red and white Burgundies originate. The northern part of the Côte d'Or is named the Côte de Nuits, after its most important (commercial) city, Nuits-Saint-Georges. This area makes red Burgundies almost exclusively, although one superb white Burgundy, Musigny Blanc, and a couple other white Burgundies do exist on the Côte de Nuits.

The southern part of the Côte d'Or, the Côte de Beaune, is named after its most important city, Beaune (the commercial and tourist center of the Côte d'Or). The Côte de Beaune makes both white and red Burgundies, but the whites are more renowned.

REMEMBER

All these red wines are entirely Pinot Noir, and the whites are entirely Chardonnay. The different characteristics from one wine to the next are due to the wines' individual terroirs.

Red Burgundy is a particularly good wine to choose in restaurants. Unlike Bordeaux and other Cabernet Sauvignon-based wines, red Burgundy is usually approachable when young because of its softness and its enticing aromas and flavours of red fruits. Moreover, red Burgundy, like all Pinot Noirs, is a versatile companion to food. It's the one red wine that can complement fish or seafood; it is ideal with salmon, for example. Chicken, turkey, and ham are also good matches for Burgundy. With richer red Burgundies, beef and game (such as duck, pheasant, rabbit, or venison) all go well.

Red Burgundy is at its best when served at cool temperatures – about 17°C. It should *not* be decanted. Even older Burgundies seldom develop much sediment, and too much aeration would cause you to lose the wonderful Burgundy aroma, which is one of the greatest features of this wine.

On the other hand, white Burgundy often benefits from decanting, especially grand cru and premier cru white Burgundies from younger vintages (five years old or younger). Great young white Burgundies, such as Corton Charlemagne, just don't evolve completely in their first few years; the extra aeration will help bring out their aromas and flavours. And remember, don't serve them too cold! The ideal temperature range for serving the better white Burgundies is 15° to 17°C.

TIP

Chablis is an ideal companion to seafood, especially oysters. Like all other white Burgundies, Chablis should be served cool (58° to 60°F, or 15°C), not cold.

The taste of fine white Burgundy

White Burgundy combines a richness of flavour – peaches, hazelnuts, and honey in Meursault; floweriness and butterscotch in a Puligny or Chassagne-Montrachet – with lively acidity and a touch of oak. With age, even more flavour complexity develops. The wine leaves a lingering reminder of all its flavours. Chardonnay wines from other regions and countries can be good, but there's nothing else quite like a great white Burgundy.

TIP

Côte Chalonnaise: Bargain Burgundies

The sad fact about Burgundy is that many of its best wines are costly. But one of Burgundy's best-kept secrets is the Côte Chalonnaise (the district that lies directly south of the Côte d'Or). True, Côte Chalonnaise Burgundies aren't as fine as Côte d'Or Burgundies (they're a bit earthier and coarser in flavour and texture), but they can still be satisfying wines — and they're about £12 to £22 retail per bottle. Four villages or communes whose names appear as appellations on wine labels are the following:

- **Mercurey *(mer cure ay)*:** Mostly red wine, and a small amount of white; the best wines of the Chalonnaise come from here, and also the most expensive (£15 to £22); three of the best producers of Mercurey are Aubert de Villaine, J.Faiveley, and Antonin Rodet.
- **Rully *(rue yee)*:** Approximately equal amounts of red and white wine; the whites, although a bit earthy, are significantly better than the reds; look for the wines of the producer Antonin Rodet.
- **Givry *(gee vree)*:** Mostly red wine, and a small amount of white; reds are better than the whites (but quite earthy); Domaine Joblot's Givry is especially worth seeking out.
- **Montagny *(mon tah nyee)*:** All white wine; look for Antonin Rodet's and Louis Latour's Montagny.

Chablis: Unique White Wines

The village of Chablis, northwest of the Côte d'Or, is the closest Burgundian commune to Paris (about a two-hour drive). Although Chablis' wines are 100 per cent Chardonnay just like the white Burgundies of the Côte d'Or, they're quite different in style. For one thing, almost all Côte d'Or white Burgundies ferment and age in oak barrels, but many Chablis producers use stainless steel tanks instead, at least for some of their wines. Also, Chablis' climate is cooler, producing wines that are intrinsically lighter-bodied, relatively austere in flavour, and crisper.

Chablis wine is classically very dry and sometimes has flinty flavours, without quite the richness and ripeness of Côte d'Or white Burgundies. (Recent Chablis vintages – such as 1997, 2000, and 2003 – have been so warm, however, that the wines have been tasting riper than usual.) For a classic, cool-climate Chablis, try a bottle from the 2004, 2002, or 1996 vintage. The 2005 Chablis vintage is supposed to be fantastic!

Chablis is at its best at the premier cru and grand cru level. Simple village Chablis is less expensive — about £11 to £17 — but you can often find better white wines from Mâcon, the Chalonnaise, or the Côte d'Or (Bourgogne Blanc) at that price.

Mâcon: Affordable Whites

The Mâconnais lies directly south of the Chalonnaise and north of Beaujolais. It has a milder, sunnier climate than the Côte d'Or to the north. Wine production centres around the city of Mâcon, a gateway city to Provence and the Riviera. The hills in the Mâconnais contain the same chalky limestone beloved by Chardonnay that can be found in many Burgundy districts to the north. In Mâcon, you can even find a village called Chardonnay.

Mâcon's white wines, in fact, are 100 per cent Chardonnay. Most of them are simply called *Mâcon* or *Mâcon-Villages* (a slightly better wine than Mâcon, because it comes from specific villages), and they retail for £7 to £10 a bottle. Often better are Mâcons that come from just one village; in those wines, the name of the village is appended to the district name, Mâcon (as in Mâcon-Lugny or Mâcon-Viré).

TIP

If you're thinking that £12 or more sounds like too much to spend for a bottle of white Burgundy or Chablis for everyday drinking, here's an alternative wine for you: white Mâcon. Many of the best white wine buys – not only in France, but in the world – come from the Mâconnais district.

TIP

The best Mâcon whites come from the southernmost part of the district and carry their own appellations – Pouilly-Fuissé *(pwee fwee say)* and Saint-Véran *(san ver ahn)*.

- Pouilly-Fuissé is a richer, fuller-bodied wine than a simple Mâcon, is often oaked, and is a bit more expensive (around £12 to £15; up to £28 for the best examples). To try an outstanding example of Pouilly-Fuissé, buy Château Fuissé, which, in good vintages, compares favorably with more expensive Cote d'Or white Burgundies.
- Saint-Véran, at £8 to £12, is very possibly the best-value wine in all of Mâcon. Especially fine is the Saint-Véran of Verget, who is one of the best producers of Mâconnais wines.

Mâcon whites are medium-bodied, crisp, fresh, and yet substantial wines, often with minerally flavour. They're usually unoaked. You should enjoy them while they're young, generally within three years of the vintage.

Beaujolais: As Delightful As It Is Affordable

The Beaujolais district is situated south of the Mâconnais, in the heart of one of the greatest gastronomic centres of the world; good restaurants abound in the area, as well as in the nearby city of Lyon. As a wine, Beaujolais is so famous that it stands apart from the other wines of Burgundy. It even has its own red grape, Gamay. The fact that Beaujolais is part of Burgundy is merely a technicality.

Beaujolais and *Beaujolais Supérieur* (1 per cent higher in alcohol) are the easiest Beaujolais wines. They're fresh, fruity, uncomplicated, fairly light-bodied wines that sell for £5 to £7 and are best a year or two after the vintage. They're fine wines for warm weather, when a heavier red wine would be inappropriate.

Beaujolais has its serious side, too. *Beaujolais-Villages* is a wine blended from grapes that produce fuller, more substantial wine than simple Beaujolais. It costs a dollar or two more, but can be well worth the difference.

Beaujolais that's even higher quality comes from ten specific areas in the north. The wines of these areas are known as *cru* Beaujolais, and only the name of the cru appears in large letters on the label. (The wines aren't actually named Beaujolais.) Cru Beaujolais have more depth and, in fact, need a little time to develop; some of the crus can age and improve for four or five years or more. They range in price from about £7 to £15.

TIP

If you're a white wine, white Zinfandel, or rosé wine drinker (or even a nonwine drinker!), Beaujolais may be the *ideal* first red wine to drink – a bridge, so to speak, to more serious red wines. Beaujolais wines are the fruitiest red wines in France, although they're dry. Beaujolais is truly a fun wine that's delicious and doesn't require contemplation.

Other French Wines

IN THIS ARTICLE

● *Saving money with robust red Rhônes* ● *White gems of the Loire and Alsace* ● *The Languedoc and Provence*

As the number one producer of wine, France has many varieties to choose from. In addition to the Bordeaux, Burgundy, and Beaujolais, several other wines stand out.

The Hearty Rhônes of the Valley

The Rhône *(rone)* Valley is in southeastern France, south of Beaujolais, between the city of Lyon in the north and Avignon directly south (just north of Provence). The growing season in the Rhône Valley is sunny and hot. The wines reflect the weather: The red wines are full, robust, and fairly high in alcohol. Even some of the white wines tend to be full and powerful. But the wines from the southern part of the Rhône are distinctly different from those in the northern Rhône Valley.

Generous wines of the south

Most (in fact, about 95 per cent of) Rhône wines come from the Southern Rhône, where the wines are generally inexpensive and uncomplicated. They're mainly blends of several grape varieties. The dominant grape variety in the southern Rhône is the prolific Grenache, which makes easygoing wines that are high in alcohol and low in tannin — but some blends contain significant amounts of Syrah or other varieties, which makes for somewhat gutsier wines. Almost all Côtes du Rhône wines are red.

Besides Côtes du Rhône, other southern Rhône wines to look for are

- **Côtes du Ventoux** *(vahn too)*, which is similar to, but a bit lighter than, Côtes du Rhône
- **Côtes du Rhône-Villages,** from 95 villages, making fuller and a bit more expensive wines than Côtes du Rhône; 16 of these villages are entitled to use their names on the label, such as "Cairanne – Côtes du Rhône-Villages"
- The single-village wines **Gigondas** *(jhee gon dahs)* and **Vacqueyras** *(vah keh rahs)*

TIP

For a good, reliable dry red wine that costs about £5 to £9, look no farther than the Rhône Valley's everyday red wine, Côtes du Rhône, which comes mainly from the southern part of the region. The Rhône Valley makes more serious wines – mostly red – but Côtes du Rhône is one of the best inexpensive red wines in the world.

Two interesting dry rosé wines of the southern Rhône are Tavel *(tah vel)* and Lirac *(lee rahk)*; Lirac is less well known and therefore less expensive. Both are made from the Grenache and Cinsault grapes. They can be delightful on hot, summer days or at picnics. As with most rosé wines, they are best when they're very young.

But Châteauneuf-du-Pape *(shah toe nuf doo pahp)* is the king in the Southern Rhône. Almost all Châteauneuf-du-Pape is red wine, and a blend of grapes: As many as 13 varieties can be used, but Grenache, Mourvèdre, and Syrah predominate. At its best, Châteauneuf-du-Pape is full-bodied, rich, round, and ripe. In good vintages, it will age well for 15 to 20 years. Most red Châteauneuf-du-Pape wines (a small amount of very earthy-style white Châteauneuf-du-Pape is also made) retail in the £17 to £28 price range, but the best ones can cost up £44 or more.

Noble wines of the north

The two best red wines of the entire Rhône – Côte-Rôtie *(coat roe tee)* and Hermitage *(er mee tahj)* – hail from the Northern Rhône Valley. Both are made from the noble Syrah grape (but some white Viognier grapes are sometimes used in Côte Rôtie).

Although both are rich, full-bodied wines, *Côte-Rôtie* is the more finesseful of the two. It has a wonderfully fragrant aroma – which always reminds us of green olives and raspberries – and soft, fruity flavours. In good vintages, Côte Rôtie can age for 20 years or more. Most Côte-Rôties are in the £28 to £53 price range.

Red Hermitage is clearly the most full-bodied, longest-lived Rhône wine. This complex, rich, tannic wine needs several years before it begins to develop, and it will age easily for 30 years or more in good vintages (2003, 1999, 1998, 1995, 1991, 1990, 1989, and 1988 were all excellent vintages in the Northern Rhone). The best red Hermitages sell today for £31 to £56, and a few are even over £62, although lesser Hermitages are as low as £25 to £28.

TIP

The three best producers of Hermitage are Jean-Louis Chave, Chapoutier, and Paul Jaboulet Ainé (for his top Hermitage, La Chapelle).

Jaboulet also makes a less expensive little brother to Hermitage, a Crozes-Hermitage (a separate appellation) called Domaine de Thalabert. It's as good as – if not better than – many Hermitages, can age and improve for 10 to 15 years in good vintages, and is reasonably priced at £17 to £19. It's a wine to buy.

Cornas, also made entirely from Syrah, is another fine Northern Rhône red wine. Cornas resembles Hermitage in that it is a huge, tannic wine that needs 10 to 20 years of aging. It ranges in price from £25 to £50. Two Cornas producers to look for are Domaine August Clape and Jean-Luc Colombo.

A small amount of white Hermitage is produced from the Marsanne and Roussanne grape varieties. White Hermitage is traditionally a full, heavy, earthy wine that needs eight to ten years to fully develop. Chapoutier's fine Hermitage Blanc, Chante-Alouette, however, is all Marsanne (about £50) and made in a more approachable style. The other great white Hermitage is Chave's; about £56, it's complex, rich, and almost as long-lived as his red Hermitage.

The Loire Valley: White Wine Heaven

If you're looking for white-wine alternatives to Chardonnay, discover the Loire *(l'wahr)* Valley wine region. A lot of white wines come from there, but virtually none of them are Chardonnay! For the record, you can find red wines and some dry rosés, too, in the Loire, but the region is really known for its white wines.

The Loire Valley stretches across northwest France, following the path of the Loire River from central France in the east to the Atlantic Ocean in the west. The rather cool climate, especially in the west, produces relatively light-bodied white wines.

In the eastern end of the Valley (Upper Loire), just south of Paris, are the towns of Sancerre and Pouilly-sur-Loire, located on opposite banks of the Loire River. Here, the Sauvignon Blanc grape makes lively, dry wines that have spicy, green-grass and minerally flavours. The two principal wines in this area are Sancerre *(sahn sehr)* and Pouilly-Fumé *(pwee foo may)*.

- Sancerre is the lighter, drier, and more vibrant of the two. It's perfect for summer drinking, especially with shellfish or light, freshwater fish, such as trout. Look for the Sancerres of Domaines Henri Bourgeois or Lucien Crochet.

- Pouilly-Fumé is slightly fuller than Sancerre and can have attractive gun-flint and mineral flavours. Pouilly-Fumé can be quite a fine wine when made by a good producer such as Didier Dagueneau or Ladoucette. Because of its fuller weight, Pouilly-Fumé goes well with rich fish, such as salmon, or with chicken or veal.

TIP

Most Sancerre and Pouilly-Fumé wines sell in the £12 to £22 range, but a few of the better Pouilly-Fumés can cost £31 or more. These wines are at their best when they're young; drink them within four years of the vintage.

The central Loire Valley is known for both its white and red wines. The Chenin Blanc grape makes better wine here than it does anywhere else in the world. The Anjou district produces arguably the world's greatest *dry* Chenin Blanc wine, Savennières (starts at about £12). A great dessert white wine made from Chenin Blanc, Coulée de Serrant, also comes from Anjou. Bonnezeaux and Quartz-de-Chaume are two other dessert white wine appellations from Anjou that are made from Chenin Blanc.

Near the city of Tours (where you can see beautiful châteaux of former French royalty), lies the town of Vouvray *(voo vray)*. Vouvray wines come in three styles: dry *(sec)*, medium-dry *(demi-sec)*, or sweet (called *moelleux*, pronounced *m'wah leuh*). Vouvray also can be a sparkling wine.

The best wines of Vouvray, the sweet *(moelleux)*, can be made only in vintages of unusual ripeness, which occur infrequently. These wines need several years to develop and can last almost forever, thanks to their remarkable acidity; their prices begin at about £13. Three renowned Vouvray producers are Philippe Foreau of Clos Naudain, Gaston Huet-Pinguet, and Didier Champalou.

TIP

Less expensive Vouvrays, priced at about £8 to £9, are pleasant to drink young. Even the drier versions aren't truly bone dry and are a good choice if you don't enjoy very dry wines. They go well with chicken or veal in cream sauce, spicy cuisines, or fruit and soft cheese after dinner.

The Loire Valley's best red wines also come from the Central Loire. Made mainly from Cabernet Franc, they carry the place-names of the villages the grapes come from: Chinon *(shee nohn)*, Bourgueil *(boorguh'y)*, Saint-Nicolas-de-Bourgueil *(san nee co lah deh boor guh'y)*, and Saumur-Champigny *(saw muhr shahm pee n'yee)*. They're all spicy, great-value (£7 to £22), medium-bodied reds that are famously food friendly.

Close to the Atlantic Ocean, the third wine district of the Loire Valley is the home of the Muscadet grape (also known as the Melon). The wine, also called Muscadet *(moos cah day)*, is light and very dry, with apple and mineral flavours – perfect with clams, oysters, and river fish (and, naturally, ideal for summer drinking).

TIP

The best news about Muscadet is the price. You can buy a really good Muscadet for £5 to £7. Buy the youngest one you can find because Muscadet is best within two years of the vintage; it isn't an ager.

Alsace Wines: French, Not German

Some wine drinkers confuse the wines of Alsace *(ahl zas)* with German wines. Alsace, in northeastern France, is just across the Rhine River from Germany. But Alsace wines are distinctly different from German wines, generally dry, and fuller bodied. Alsace wines are unique among French wines in that almost all of them carry a grape variety name as well as a place-name (that is, Alsace). All Alsace wines come in a long-necked bottle called a *flûte*. The wines of Alsace also happen to represent very good value.

Considering Alsace's northerly latitude, you'd expect the region's climate to be cool. But thanks to the protection of the Vosges Mountains to the west, Alsace's climate is quite sunny and temperate, and one of the driest in France – in short, perfect weather for grape growing.

Although some Pinot Noir exists, 91 per cent of Alsace's wines are white. Four are particularly important: Riesling, Pinot Blanc, Pinot Gris, and Gewurztraminer. Each reflects the characteristics of its grape, but they all share a certain aroma and flavour, sometimes called a spiciness, that can only be described as the flavour of Alsace.

Riesling is the king of Alsace wines (remember that it's a *dry* wine here). Alsace Riesling has a fruity aroma but a firm, dry, almost steely taste. Although, like most Alsace wines, it can be consumed young, a Riesling from a good vintage can easily age and improve for ten years or more. Rieslings are in the £9 to £22 price range; a few of the best are more.

TIP

Alsace **Pinot Blanc** is the lightest of the four wines. Some producers make their Pinot Blanc medium-dry to appeal to wine drinkers who are unfamiliar with the region's wines, while others make classic, bone-dry Pinot Blanc. Either way, it's best young. Pinot Blanc is quite inexpensive, selling for £7 to £12.

Pinot Gris or **Tokay-Pinot Gris** is made from Pinot Gris, the same variety that you find in Italy as Pinot Grigio. Here in Alsace, it makes a rich, spicy, full-bodied, characterful wine. Alsace's Pinot Gris retails for £9 to £18; it goes well with spicy meat dishes and can work with slightly sweet or sour flavours.

TIP

The **Gewurztraminer** grape has such intense, exotic, spicy aromas and flavours that it's a love-it-or-leave-it wine. But it certainly has its followers. And this grape is clearly at its best in Alsace. If you haven't tried an Alsace Gewurztraminer yet, you haven't tasted one of the most distinctive wines in the world. It's quite low in acidity and high in alcohol, a combination that gives an impression of fullness and softness. It goes best with *foie gras* and strong cheeses, and some people like it with spicy Asian cuisine. Gewurztraminer sells for about the same price as Riesling but doesn't age quite as well.

The South and Southwest

The most dynamic wine regions in France are all located in the southern part of the country. Ironically, this is the oldest wine-producing area in France: The South is also the part of France that makes the most wine. Languedoc-Roussillon, a dual wine region, produces over 40 per cent of France's wine!

Southwest France, the huge area between Bordeaux and the Spanish border, also makes wine, and many wine regions here have also experienced a renaissance. Like the South, it's mainly red wine country, but you can find some interesting whites, rosés, sparkling wines, and dessert wines, as well. You might say that the South and Southwestern France are the country's "new" frontiers.

The Midi: France's bargain basement

The sunny, dry Languedoc-Roussillon *(lahn guh doc roo see yohn)* region, also known as the Midi *(mee dee)*, has long been the country's largest wine-producing area. The region makes mainly red wines; in fact, more than half of France's red wines come from here. Traditionally, these robust red wines came from typical grape varieties of the South, such as Carignan, Cinsault, and Grenache. But in the last two decades, more serious varieties such as Syrah, Cabernet Sauvignon, and Merlot have become popular with growers. Winemakers use these grapes both for varietal wines and in blends.

In this region, look especially for the red wines from the AOC zones of **Corbières, Minervois, St-Chinian, Fitou**, and **Costières de Nîmes.** In addition, many varietal wines carrying the designation *Vin de Pays d'Oc* are often good value They're made from grapes that can come from anywhere in the Languedoc-Roussillon region, rather than from a specific AOC zone.

The best news is that most of these wines are in the £5 to £9 price range, although a few of the better ones cost £12 or more. Two of the better known brands of varietal wines (with the Vin de Pays d'Oc appellation) are Fortant de France and Réserve St. Martin.

Timeless Provence

Provence *(pro vahns)* – southeast of the Rhône Valley, east of Languedoc-Roussillon, and west of Northern Italy – may be France's most beautiful region. Home of the Riviera, Nice, and Cannes, it's certainly the country's most fashionable and touristy region. The excellent light and climate have always attracted great artists.

Wine has always been part of Provence's culture and economy. Provence is best known for its rosés, which so many tourists enjoy on the Riviera, but Provence's red wines are now winning the most critical acclaim. Rosé wines still dominate in the region's largest AOC wine zone, Côtes de Provence, but in three other important AOC zones – Coteaux d'Aix-en-Provence, Les Baux-de-Provence, and Bandol – red wines rule. Bandol, and its foremost producer, Domaine Tempier, enjoy Provence's greatest reputation for reds. Cassis (no relation to the blackcurrant liqueur of the same name), a small AOC zone on the Mediterranean coast near Marseilles, makes distinctive, aromatic white wines.

Provence's reds and rosés derive from the same grape varieties used in Languedoc-Roussillon – Grenache, Cinsault, Mourvèdre, Carignan, Syrah, and Cabernet Sauvignon. The main varieties in white Cassis are Clairette and Marsanne.

Southwest France

The large area that borders the Atlantic Ocean south of the Bordeaux region is known as Southwest France – but it's actually composed of many individual wine districts. Three of the most significant are situated near Bordeaux.

- **Bergerac** *(ber jhe rak)* makes Bordeaux-like red and white wines, without the Bordeaux prices. Merlot dominates Bergerac's red wines, while Sémillon and Sauvignon Blanc are the main varieties for its whites, some of which cost as little as £5 a bottle.
- **Monbazillac** *(mon bah zee yak)* specializes in sweet dessert wines similar to the Sauternes of the Bordeaux region, but Monbazillac's wines are considerably less expensive – and less complex – than Sauternes.
- **Cahors** *(cah or)* is Southwest France's most prestigious red wine district. The main grape variety is Malbec, and that name increasingly appears on the labels. Nowhere else in the world, except Argentina, does this variety play such an important role. The best wines of the traditional Cahors producers, such as Château Lagrezette, are dark, tannic reds that need about ten years of aging before they mature. Prices for Château Lagrezette Cahors wines begin at about £14.

Wine in Italy: Piedmont and Tuscany

IN THIS ARTICLE

- *Getting familiar with Italy's wine zones*
- *Looking at Piedmont's reds*
- *Relaxing in Tuscany*

Thanks to the popularity of Italian restaurants, most people have frequent opportunities to enjoy best-selling Italian wines, such as Pinot Grigio, Soave, Valpolicella, and Chianti. But Italy makes other wines, too – many of them among the greatest wines on earth. And just about every one of Italy's thousand-something wines is terrific with food, because Italian wines are made specifically to be enjoyed during a meal. That's how the Italians drink them.

This article focuses on two of Italy's three most renowned wine areas – Piedmont and Tuscany. The next article covers the wines of Northeastern Italy.

The Vineyard of Europe

Tiny, overachieving Italy – 60 per cent the size of France, three-quarters the size of California – makes more than 20 per cent of the world's wines! Wine is the lifeblood of the Italian people. Vines grow all over, and no meal can possibly occur without a bottle of wine on the table.

The downside to wine's penetration into Italian culture is that Italians often take wine for granted. Italy has yet to incorporate official recognition of her best vineyard sites *(crus)* into her wine laws, as the French have done in Burgundy. Italy's casual attitude toward wine slowed the acceptance of even the top wines by many serious wine lovers around the world – although recognition of Italian wines has grown considerably over the past decade.

Another handicap of Italian wines, for wine drinkers in other countries who want to learn about them, is that most Italian wines are made from native grape varieties that don't exist elsewhere (and when transplanted, don't perform nearly as well as in Italy). Grapes such as Nebbiolo, Sangiovese, Aglianico, and Barbera, to name just a few, can make outstanding wine in Italy, but their names are unfamiliar.

On the upside, Italy is blessed with such a variety of soils and climates – from Alpine foothills in the north to Mediterranean coastlines – that the range of her wines is almost endless. Italy's hilly landscape provides plenty of high-altitude relief for grapevines even in the warm south.

The ordinary and the elite

Italy's wines, as wine lovers outside of Italy know them, fall into two distinct groups:

- Inexpensive red and white wines for everyday drinking with meals in the casual Italian fashion
- The better wines, which range from good to great in quality

TIP: One of the best-known Italian wines in the first category is Pinot Grigio. In the second category is Barolo, one of the world's finest red wines, along with many other fine Italian wines.

Italy's wine regions

Italy is said to have 20 wine regions, which correspond exactly to her political regions (see figure). In other words, wine is produced everywhere in Italy.

Many of the finest wines come from the north: the Piedmont region in the northwest, Tuscany in north-central Italy, and the three regions (informally called the Tre Venezie) of Northeastern Italy.

Reds Reign in Piedmont

Piedmont's claim to wine fame is the Nebbiolo *(neb bee OH lo)* grape, a noble red variety that produces great wine only in northwestern Italy. The proof of Nebbiolo's nobility is its wines: Barolo *(bah RO lo)* and Barbaresco *(bar bah RES co)* are two of the world's great red wines. Both are wines made entirely from the Nebbiolo grape in the Langhe hills around Alba, and each is named after a village within its production zone.

The wine regions of Italy.

TIP

Most Barolos and Barbarescos aren't wines to drink young. Production rules stipulate that Barolo isn't Barolo until it has aged for three years at the winery, or for five years if it is called *Riserva*. (Barbaresco's minimum aging is two years, or four for Riserva.) But both wines benefit from additional aging. When traditionally made, Barolo and Barbaresco often require 10 to 15 years' total aging, from the year of the vintage – and they usually benefit from a few hours of aeration before drinking to soften their somewhat tough tannins.

Both Barolo and Barbaresco are robust reds – very dry, full-bodied, and high in tannin, acidity, and alcohol. Their aromas suggest tar, violets, roses, ripe strawberries, and (sometimes) truffles – the kind that grow in the ground, not the chocolate! Barolo is more full-bodied than Barbaresco and usually requires a bit more aging; otherwise, the two wines are very similar. Like most Italian wines, they show at their best with food. Good Barolo and Barbaresco wines usually start at £25 and run to well over £63 per bottle.

Both Barbaresco and especially Barolo have something in common with Burgundy in France: *You must find a good producer to experience the wine at its best.*

REMEMBER

What is called a wine region in France, such as Burgundy or Alsace, is referred to as a wine *zone* in Italy to avoid confusion with the political region.

Some producers – including Giacomo Conterno, both Mascarellos, Giuseppe Rinaldi, and Bruno Giacosa – clearly make traditionally styled wines; others – such as Gaja, Altare, and Clerico – make modern-style wines; and some, such as Ceretto and Vietti, combine aspects of both winemaking styles.

Weekday reds

The Piedmontese reserve serious wines like Barolo and Barbaresco for Sunday dinner or special occasions. What they drink on an everyday basis are the red wines Dolcetto *(dohl CHET to)*, Barbera *(bar BEAR ah)*, and Nebbiolo (grown outside of prestigious DOCG zones such as Barolo and Barbaresco). Of the three, Dolcetto is the lightest-bodied and is usually the first red wine served in a Piedmontese meal.

Dolcetto

If you know enough Italian to translate the phrase *la dolce vita,* you may think that the name Dolcetto indicates a sweet wine. Actually, the Dolcetto grape tastes sweet but the *wine* is distinctly dry and somewhat grapey with noticeable tannin. Dolcetto is often compared to Beaujolais (France's easy-drinking red wine), but it's drier and more tannic than most Beaujolais wines and goes better with food.

Dolcetto sells for £6 to £15. The best Dolcetto wines are from the zones of Dogliani, Diano d'Alba, and Alba; the labels of these wines carry the grape name, Dolcetto, along with the name of the area. Just about all recommended Barolo producers make a Dolcetto, usually Dolcetto *d'Alba* (from Alba). A favourite producer who happens to make only Dolcetto di Dogliani is Quinto Chionetti.

Barbera

While Dolcetto is unique to Piedmont, the Barbera grape is the second most widely planted red grape variety in all of Italy. (Sangiovese is *the* most planted red variety.) But it's in Piedmont – specifically the Asti and Alba wine zones – that Barbera excels. It's a rich, red wine with high acidity and generous black-cherry fruit character.

Barbera d'Alba is generally a bit fuller, riper, and richer than the leaner Barbera d'Asti – but Barbera d'Asti from certain old vineyards rivals Barbera d'Alba in richness and in power. (Link the *d'* with the word following it when pronouncing these names: *DAL ba, DAHS tee.*) Barbera happens to be a favourite everyday wine, especially with pasta or pizza – or anything tomatoe-y.

Two different styles of Barbera are available:

- The traditional style, aged in *casks* (large oak containers that impart little, if any, oak flavour to the wine), which sells for about £7 to £15.
- The newer style, oak-influenced Barbera aged in *barriques* (small, new barrels of French oak), which sells in the £15 to £28 range

Both types of Barbera are very good; with a few exceptions, you may prefer the simpler, less expensive, traditional style. Barbera is an unusual red grape variety in that it has practically no tannin, and so the tannins from the small oak barrels can complement Barbera wines.

TIP

Two particularly good producers of Barbera d'Alba are Vietti and Giacomo Conterno. Vietti also makes a terrific Barbera d'Asti called "La Crena."

Nebbiolo

A third weekday red from Piedmont is Nebbiolo (d'Alba or Langhe), made from Nebbiolo grapes grown in vineyards outside the prized Barolo or Barbaresco zones. The wine is lighter in body and easier to drink than either Barolo or Barbaresco, and it sells for about £9 to £12 a bottle. Another variation is Roero Rosso, made almost entirely from Nebbiolo.

Whites in a supporting role

Almost all Piedmont's wines are red, but the region does boast two interesting dry whites. Gavi, named for a town in southern Piedmont, is a very dry wine with pronounced acidity, made from the Cortese grape. Most Gavis sell for £8 to £15, while a premium Gavi, La Scolca's Black Label, costs around £28.

Arneis *(ahr NASE)* is a white wine produced in the Roero zone near Alba from a long-forgotten grape also called Arneis, which was rescued by Alfredo Currado, owner of Vietti winery, nearly 40 years ago. Arneis is a dry to medium-dry wine with rich texture. It's best when it is consumed within two years of the vintage; a bottle sells for £12 to £17. Besides Vietti's, look for Bruno Giacosa's and Ceretto's Arneis.

Tuscany's Chianti: Italy's Great, Underrated Red

Chianti is a large wine zone extending through much of Tuscany. The zone has eight districts. Chianti wines may use the name of the district where their grapes grow or the simpler appellation, Chianti, if their production doesn't qualify for a district name (if grapes from two districts are blended, for example).

The district known as *Chianti Classico* is the heartland of the zone, the best area, and the one district whose wines are widely available. The only other Chianti district that comes close to rivaling Chianti Classico in quality is *Chianti Rufina* (*ROO fee nah;* not to be confused with the renowned Chianti producer Ruffino), whose wines are fairly available, especially from the well-known producer Frescobaldi.

Besides varying according to their district of production, Chianti wines vary in style according to their aging. *Riserva* wines must age for two years or more at the winery, and some of this aging is often in French oak; the best riservas have potential for long life. Chianti wines can also vary slightly according to their grape blend: Many top Chiantis are made almost entirely from the Sangiovese grape, while others use up to 25 per cent of other varieties, including "international" varieties, such as Cabernet Sauvignon, Merlot, and Syrah.

Chianti is a very dry red wine that, like most Italian wines, tastes best with food. It ranges from light-bodied to almost full-bodied, according to the district, producer, vintage, and aging regime. It often has an aroma of cherries and sometimes violets and has a flavour reminiscent of tart cherries. The best Chianti wines have very concentrated fruit character and usually taste best from five to eight years after the vintage – although in good vintages they have no problem aging for ten or more years.

TIP

The two exceptional vintages to look for in Chianti wines are 1999 and 2001 — two of the better Tuscan vintages of modern times.

Although Chianti isn't a huge wine, today's Chianti wines, especially Chianti Classico, are richer and more concentrated than ever before — in some cases, too rich. Recent warm vintages in Europe, such as 1997, 2000, and 2003, have fed a trend toward ripeness, fleshiness of texture, and higher alcohol. The addition of international varieties and the use of barriques for aging — especially for riservas — has also affected the wines. More than ever, you must choose your Chianti producers with care.

From simple £6 to £9 Chianti to the more substantial Chianti Classico (generally between £9 and £15), Chianti remains one of the wine world's great values. Chianti Classico Riservas are a bit more costly, ranging from £17 to £28 per bottle.

Vino Nobile, Carmignano, and Vernaccia

Three more Tuscan wines of note include two reds – Vino Nobile di Montepulciano *(NO be lay dee mon tay pul chee AH no)* and Carmignano *(car mee NYAH no)* – and Tuscany's most renowned white wine, Vernaccia di San Gimignano *(ver NAH cha dee san gee mee NYAH no)*.

Monumental Brunello di Montalcino

While Chianti has been famous for centuries, another great Tuscan wine, Brunello di Montalcino, exploded on the international scene only some 35 years ago. Today, Brunello di Montalcino *(brew NEL lo dee mon tahl CHEE no)* is considered one of the greatest, long-lived red wines in existence. It has a price-tag to match: £28 to over £126 a bottle (for wines by the producer Soldera).

The wine is an intensely concentrated, tannic wine that demands aging (up to 20 years) when traditionally made, and benefits from several hours of aeration before serving. Lately, some producers in Montalcino have been making a more approachable style of Brunello.

Rosso di Montalcino is a less expensive (£14 to £18), readier-to-drink wine made from the same grape and grown in the same production area as Brunello di Montalcino. Rosso di Montalcino from a good Brunello producer is a great value, offering you a glimpse of Brunello's majesty without breaking the bank.

Traditional winemakers, such as Biondi-Santi, Soldera, Costanti, Canalicchio di Sopra, and Pertimali, make wines that need at least 15 to 20 years of aging in good vintages (2001, 1999, 1997, 1995, 1990, 1988, 1985, and 1975 are recent great vintages for Brunello). Brunellos from avowed modern-style producers, such as Caparzo, Altesino, and Col d'Orcia, can be enjoyed in ten years. Younger than ten years – drink Rosso di Montalcino.

TIP

The Montepulciano wine zone, named after the town of Montepulciano, is southeast of the Chianti zone. Vino Nobile's principal grape is the Prugnolo Gentile (a.k.a. Sangiovese). From a good producer, Vino Nobile di Montepulciano can rival the better Chianti Classicos.

The Carmignano wine region is directly west of Florence. Although Sangiovese is the main grape of Carmignano – just as it is for Chianti – Cabernet Sauvignon is also one of this wine zone's traditional grapes. As a result, Carmignano's taste is rather akin to that of a Chianti with the finesseful touch of a Bordeaux. Two outstanding producers of Carmignano are Villa di Capezzana and Ambra.

Vernaccia di San Gimignano is named for the medieval walled town of San Gimignano, west of the Chianti Classico zone. Vernaccia is generally a fresh white wine with a slightly oily texture and an almond flavour, and it is meant to be drunk young. For an unusual interpretation, try Teruzzi & Puthod's oak-aged riserva, Terre di Tufo, a pricey but very good Vernaccia (about £11). Most Vernaccias are in the £6 to £8 range. Besides Teruzzi & Puthod, producers to seek are Montenidoli, Mormoraia, Cecchi, and Casale-Falchini.

Two more reds and a white

So many good Tuscan wines exist that three more deserve a mention here.

Pomino (*po MEE no*) is the name of a red and a white wine from a tiny area of the same name, which lies within the Chianti Rufina district. A hilly land with a particularly mild climate, Pomino has long been a stronghold of French varieties, right in the heart of Sangiovese-land. The Frescobaldi family, the major producer of Chianti Rufina, is also the main landowner in Pomino, and makes both a Pomino Rosso and a Pomino Bianco—but the red is the more noteworthy of the two. Pomino Rosso is a blend of Cabernet Sauvignon, Cabernet Franc, Merlot, Sangiovese, Canaiolo — and in Frescobaldi's case, also Pinot Nero (the Italian name for Pinot Noir). It sounds like a crazy blend, but it works! Pomino Bianco blends Chardonnay, Pinot Bianco, Trebbiano, and often Pinot Grigio. Besides Frescobaldi, another fine producer of Pomino is Fattoria Petrognano. Pomino costs about £15 to £18.

The new frontier for Tuscan wine is the Maremma (mah REM mah) in southwest Tuscany. Of the many wines coming from this area, two that impress us are the red, Morellino di Scansano, and the white, Vermentino. Morellino (moh rehl LEE no) is the name for Sangiovese in the hilly area around the town of Scansano (scahn SAH no). Most Morellino di Scansano wines offer an easy-drinking, inexpensive (£7 to £12) alternative to Chianti —although a few high-end examples exist. Look for Fattoria Le Pupille and Moris Farms.

Vermentino, an aromatic white varietal wine popular in Sardinia and Liguria, has suddenly become the hot new variety in Tuscany, especially along the coast. It's a crisp, medium-bodied, flavourful white that's usually unoaked, and sells for £9 to £12. Many leading Tuscan producers, such as Antinori and Cecchi, are making attractive Vermentino wines.

Tre Venezie and Other Italian Regions

IN THIS ARTICLE

- *Exploring the northeastern corner of Italy*
- *Finding fine wines from other parts of Italy*

More than 2,000 years after Julius Caesar conquered Gaul, the Italians continue to take the world by storm. With passion, artistic flair, impeccable taste, and flawless workmanship as their weapons, the Italians have infiltrated the arenas of fashion, film, food, and, of course, wine.

While Piedmont and Tuscany are famous in their own rights, they aren't the only Italian zones that produce fine wines. In fact, the whole country of Italy is known for its wines. This article discusses some of the other Italian regions whose wines you're likely to find in your wine shop or *ristorante.*

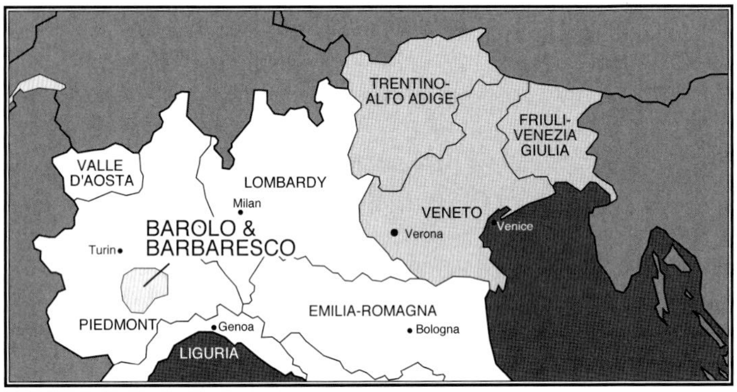

Tre Venezie

The three regions in the northeastern corner of Italy are often referred to as the *Tre Venezie* – the Three Venices – because they were once part of the Venetian Empire. Colourful historical associations aside, each of these regions produces red and white wines that are among the most popular Italian wines outside of Italy – as well as at home.

Three gentle wines from Verona

Chances are that if your first dry Italian wine wasn't Chianti or Pinot Grigio, it was one of Verona's big three: the white **Soave** *(so AH vay)* or the reds, **Valpolicella** *(val po lee CHEL lah)* or **Bardolino** *(bar do LEE noh).* These enormously popular wines hail from Northeast Italy, around the picturesque city of Verona – Romeo and Juliet's hometown – and the beautiful Lake Garda.

Of Verona's two reds, Valpolicella has more body. (Bolla and Masi are two of the largest producers.) The lighter Bardolino is a pleasant summer wine when served slightly cool. Soave can be a fairly neutral-tasting unoaked white or a characterful wine with fruity and nutty flavour, depending on the producer.

TIP

Most Valpolicella, Bardolino, and Soave wines are priced from £5 to £8, as are two other white wines of the region, Bianco di Custoza and Lugana. Some of the better Veronese wines, from the following recommended producers, have slightly higher prices:

- **Soave:** Pieropan, Inama, Gini, Santa Sofia
- **Valpolicella:** Allegrini, Quintarelli, Dal Forno, Le Ragose, Bertani, Alighieri, Tommasi, Masi
- **Bardolino:** Guerrieri-Rizzardi, Cavalchina, Fratelli Zeni

Amarone della Valpolicella (also simply known as Amarone), one of Italy's most full-bodied red wines, is a variant of Valpolicella. It's made from the same grape varieties, but the ripe grapes dry on mats for several months before fermentation, thus concentrating their sugar and flavours. The resulting wine is a rich, potent (14 to 16 per cent alcohol), long-lasting wine, perfect for a cold winter night and a plate of mature cheeses. Some of the best producers of Amarone are Quintarelli, Bertani, Masi, Tommasi, Le Ragose, Allegrini, and Dal Forno.

The Austrian-Italian alliance

If you've travelled much in Italy, you probably realize that in spirit Italy isn't one unified country but 20 or more different countries linked together politically. Consider Trentino-Alto Adige. Not only is this mountainous region (the northernmost in Italy) dramatically different from the rest of Italy, but also the mainly German-speaking Alto Adige (or South Tyrol) in the north is completely different from the Italian-speaking Trentino in the south. (Before World War I, the South Tyrol was part of the Austro-Hungarian Empire.) The wines of the two areas are different, too – yet the area is considered a single region!

Alto Adige produces red wine, but most of it goes to Germany, Austria, and Switzerland. The rest of the world sees Alto Adige's white wines – Pinot Grigio, Chardonnay, Pinot Bianco, Sauvignon, and Gewürztraminer – which are priced mainly in the £7 to £11 range.

TIP

One local red wine to seek out is Alto Adige's **Lagrein**, from a native grape variety of the same name. It's a robust, hearty wine, somewhat spicy and rustic in style, but it offers a completely unique taste experience. Hofstätter and Alois Lageder are two producers who make a particularly good Lagrein.

Alto Adige produces Italy's best white wines, along with nearby Friuli. Four producers to look for are Alois Lageder, Hofstätter, Tiefenbrunner, and Peter Zemmer. Here are some highlights of each brand:

- Lageder's Pinot Bianco from the Haberlehof vineyard and his Sauvignon from the Lehenhof vineyard are exceptional examples of their grape varieties and are among the best wines from these two varieties that we've tasted.
- Hofstätter's Gewürztraminer, from the Kolbenhof vineyard, is as fine a wine as you can find from this tricky grape variety. Hofstätter also makes one of Italy's best Pinot Nero wines, Villa Barthenau.
- Tiefenbrunner's Müller-Thurgau *(MOOL lair TOOR gow)* from his Feldmarschall Vineyard (the region's highest in altitude) could well be the wine world's best wine from this otherwise lacklustre variety.
- Peter Zemmer produces reliable Chardonnay and Pinot Grigio wines in the £7 to £8 price range.

Trentino, the southern part of the Trentino-Alto Adige region, is not without its own notable wines. Some excellent Chardonnay wines come from Trentino, for example; two of the best are made by Pojer & Sandri and Roberto Zeni. (In fact, we recommend any of the wines from these two producers.) Elisabetta Foradori is a Trentino producer who specializes in red wines made from the local variety, Teroldego *(teh ROLL day go)* Rotaliano. Her best red wines, Granato and Sgarzon, are based on Teroldego and always get rave reviews from wine critics. Also, one of Italy's leading sparkling wine producers, Ferrari, is in Trentino.

The far side: Friuli-Venezia Giulia

Italy has justifiably been known in the wine world for its red wines. But in the past 20 years, the region of Friuli-Venezia Giulia, led by the pioneering winemaker, the late Mario Schiopetto, has made the world conscious of Italy's white wines as well.

Near the region's eastern border with Slovenia, the districts of Collio and Colli Orientali del Friuli produce Friuli's best wines. Red wines exist here, but the white wines have given these zones their renown. In addition to Pinot Grigio, Pinot Bianco, Chardonnay, and Sauvignon, two local favourites are Tocai Friulano and Ribolla Gialla (both fairly rich, full, and viscous).

TIP

A truly great white wine made here is Silvio Jermann's Vintage Tunina, a blend of five varieties, including Pinot Bianco, Sauvignon, and Chardonnay. Vintage Tunina is a rich, full-bodied, long-lived white of world-class status. It sells in the £20 to £50 range and, frankly, it's worth the money. Give the wine about ten years to age and then try it with rich poultry dishes or pasta.

Here's a list of recommended producers in Friuli alphabetically:

Great Producers in Friuli

Abbazia di Rosazzo/Walter Filiputti	Ronco del Gelso
Borgo Conventi	Ronco del Gnemiz
Girolamo Dorigo	Ronco dei Rosetti, of Zamò
Livio Felluga	Ronco dei Tassi
Gravner	Russiz Superiore, of Marco Felluga
Jermann	Mario Schiopetto
Miani	Venica & Venica
Lis Neris-Pecorari	Vie di Romans
Plozner	Villa Russiz
Doro Princic	Volpe Pasini

Snapshots from the Rest of Italy

Italy's wines are by no means confined to the five regions individually discussed in this and the preceding two chapters. A quick tour of some of Italy's other regions proves the point:

- **Lombardy:** In the northern part of this northerly region, near the Swiss border, the Valtellina wine district produces four relatively light-bodied red wines from the Nebbiolo grape: Sassella, Inferno, Grumello, and Valgella. Most of these wines are inexpensive (about £6 to £12) and, unlike Barolo or Barbaresco, can be enjoyed young. Lombardy is also the home of Italy's best sparkling wine district, Franciacorta.
- **Emilia-Romagna:** This is the home of Lambrusco, one of Italy's most successful wines on export markets. For a different Lambrusco experience, try a dry one if you can find it. (You may have to go to Emilia-Romagna for that – but, hey, that's not so bad. Bologna and Parma, two gastronomic meccas, are in this region.)
- **Liguria:** This narrow region south of Piedmont, along the Italian Riviera, is also the home of Cinque Terre, one of Italy's most picturesque areas. The region's two fine white wines, Vermentino and Pigato, are just made for Liguria's pasta with pesto, its signature dish.
- **Marches** (also known as **Marche**): Verdicchio is a dry, inexpensive white wine that goes well with fish, is widely available, and improves in quality with every vintage. Try the Verdicchio dei Castelli di Jesi from Fazi-Battaglia, Colonnara, or Umani Ronchi, great value at £5 to £7. Marche's best red wine, Rosso Cònero, at £9 to £13, is one of Italy's fine red wine buys.
- **Umbria:** This region, home to the towns of Perugia and Assisi, makes some good reds and whites. Orvieto, a white, is widely available for around £6 from Tuscan producers such as Antinori and Ruffino. Two interesting red wines are Torgiano, a Chianti-like blend (try Lungarotti's Rubesco Riserva DOCG), and Sagrantino di Montefalco DOCG, a medium-bodied, characterful wine made from the local Sagrantino grape.
- **Latium:** This region around Rome makes the ubiquitous, inexpensive Frascati, a light, neutral wine from the Trebbiano grape; Fontana Candida is a popular brand.
- **Abruzzo:** Montepulciano d'Abruzzo, an inexpensive, easy-drinking, low tannin, low-acid red wine, comes from here; it's a terrific everyday red, especially from a quality producer such as Masciarelli. Abruzzo is also home to two other fine producers, Cataldi Madonna and the late, great Eduardo Valentini, whose sought-after Trebbiano d'Abruzzo is perhaps the world's greatest white wine from the otherwise ordinary Trebbiano grape.
- **Campania:** Some of the best wines in Southern Italy are produced here, around Naples. The full-bodied, tannic Taurasi, a DOCG wine from the Aglianico grape, is one of the great, long-lived red wines in Italy. Premium producers are Mastroberardino (look for his single-vineyard Taurasi, called Radici), Feudi di San Gregorio, and Terredora. The same producers also make two unique whites, Greco di Tufo and Fiano di Avellino. They're full-flavoured, viscous wines with great aging capacity that sell in the £11 to £15 range. Falanghina (£7 to £10) is another exciting, light-bodied white Campania wine.
- **Basilicata:** The instep of the Italian boot, Basilicata has one important red wine, Aglianico del Vulture. It's similar to Taurasi, but not quite so intense and full-bodied. D'Angelo and Paternoster are leading producers.

- **Apulia:** This region makes more wine than any other in Italy. Generally, it is inexpensive, full-bodied red wine, such as Salice Salentino (from the native variety, Negroamaro) and Primitivo.

- **Sicily:** Once known only for its Marsala, a sweet, fortified wine, Sicily is now making quality reds and whites. Established wineries such as Corvo (a.k.a. Duca di Salaparuta) and Regaleali have been joined by exciting, new wineries such as Planeta, Morgante, Donnafugata, and Benanti to produce some of Italy's more intriguing wines – especially reds, many made from Sicily's superb variety, Nero d'Avola.

- **Sardinia:** This large island off the eastern coast of Italy makes delicate white wines and characterful reds from native grape varieties and from international varieties such as Cabernet Sauvignon. Sella & Mosca, Argiolas, and Santadi are three leading producers. Two of the more popular Sardinian wines are the white Vermentino and the red Cannonau (the local version of Grenache), both of which sell in the £6 to £10 range.

Vintages always matter/vintages don't matter

The difference between one vintage and the next of the same wine is the difference between the weather in the vineyards from one year to the next (barring extenuating circumstances such as replanting of the vineyard, new ownership of the winery, or the hiring of a new winemaker). The degree of vintage variation is thus equivalent to the degree of weather variation.

In some parts of the world, the weather varies a lot from year to year, and for wines from those regions, vintages certainly do matter. In Bordeaux, Burgundy, Germany, and most of Italy, for example, weather problems (frost, hail, ill-timed rain, or insufficient heat) can affect one vintage for the worse, while the next year may have no such problems. Where a lot of weather variation exists, the quality of the wine can swing from mediocre to outstanding from one year to the next.

In places where the weather is more predictable year after year (like much of California, Australia, and South Africa), vintages can still vary, but the swing is narrower. Serious wine lovers who care about the intimate details of the wines they drink will find the differences meaningful, but most people won't.

Another exception to the "Vintages always matter" myth is inexpensive wine. Top-selling wines that are produced in large volume are usually blended from many vineyards in a fairly large area. Swings in quality from year to year are not significant.

Does the Glass Really Matter?

If you're just drinking wine as refreshment with your meal, and you aren't thinking about the wine much as it goes down, the glass you use probably doesn't matter too much.

But if you have a good wine, a special occasion, friends who want to talk about the wine with you, or the boss for dinner, *stemware* (glasses with stems) is called for. And it's not just a question of etiquette and status: Good wine tastes better out of good glasses.

Think of wine glasses as being like stereo speakers. Any old speaker brings the music to your ears, just like any old glass brings the wine to your lips. But you can appreciate the sound so much more from good speakers. The same principle holds true with wine and wine glasses. You can appreciate wine's aroma and flavour complexities so much more out of a fine wine glass. The medium is the message.

Choose thin but not tiny

Believe it or not, the taste of a wine changes when you drink the wine out of different types of glasses. Three aspects of a glass are important: its size, its shape, and the thickness of the glass.

- **Size:** For dry red and white wine, small glasses are all wrong. Small glasses can work adequately only for sherry or dessert wines, which have strong aromas to begin with and are generally consumed in smaller quantities than table wines. In most cases, larger is usually better.
- **Thickness:** Stemware made of very thin, fine crystal costs a lot more than normal glasses. That's one reason why many people don't use it, and why some people do. The better reason for using fine crystal is that the wine tastes better out of it.
- **Shape:** The shape of the bowl also matters. Some wine glasses have very round bowls, while others have more elongated, somewhat narrower bowls. When you're having dinner at home, try your wine in glasses of different shapes to see which glass works best for that wine.

TIP

Glasses for red wines should hold a minimum of 350 ml; many of the best glasses have capacities ranging from 450 to 700 ml, or more. For white wines, 300 to 250 ml should be the minimum capacity. For sparkling wines, 250 to 350 ml capacity is fine.

Tulips, flutes, trumpets, and other wine glass names

A tulip and a flute are types of glasses designed for use with sparkling wine. The tulip is the ideally shaped glass for Champagne and other sparkling wines (see figure). It is tall, elongated, and narrower at the rim than in the middle of the bowl. This shape helps hold the bubbles in the wine longer, not allowing them to escape freely (the way the wide-mouthed, so-called Champagne glasses do).

The flute is another good sparkling wine glass (see figure), but it is less ideal than the tulip because it does not narrow at the mouth. The trumpet actually widens at the mouth, making it less suitable for sparkling wine but very elegant looking (see figure). Another drawback of the trumpet glass is that, depending on the design, the wine can actually fill the whole stem, which means the wine warms up from the heat of your hand as you hold the stem. Avoid the trumpet glass.

An oval-shaped bowl that is narrow at its mouth (see figure) is ideal for many red wines, such as Bordeaux, Cabernet Sauvignons, Merlots, Chiantis, and Zinfandels. On the other hand, some red wines, such as Burgundies, Pinot Noirs, and Barolos, are best appreciated in wider-bowled, apple-shaped glasses (see figure). Which shape and size works best for which wine has to do with issues such as how the glass's shape controls the flow of wine onto your tongue.

TIP

Good wine glasses are always clear. Those pretty pink or green glasses may look nice in your china cabinet, but they interfere with your ability to distinguish the true colors of the wine.

Glasses for sparkling wine (from left): tulip, flute, trumpet.

The Bordeaux glass (left) and the Burgundy glass.

Intriguing Wines from Old Spain

IN THIS ARTICLE

- *Looking at Spain's wine regions*
- *Ruling the roost with Rioja*
- *Getting acquainted with Ribera del Duero and Priorato*

Spain is a hot, dry, mountainous country with more vineyard land than any other nation on earth. It ranks third in the world in wine production, after France and Italy.

Spanish wine has awakened from a long period of dormancy and underachievement. Spain is now one of the wine world's most vibrant arenas. For decades, only Spain's most famous red wine region, Rioja *(ree OH ha)*, and the classic fortified wine region, Sherry, had any international presence for fine wines. Now, many other wine regions in Spain are making seriously good wines.

The wine regions of Spain.

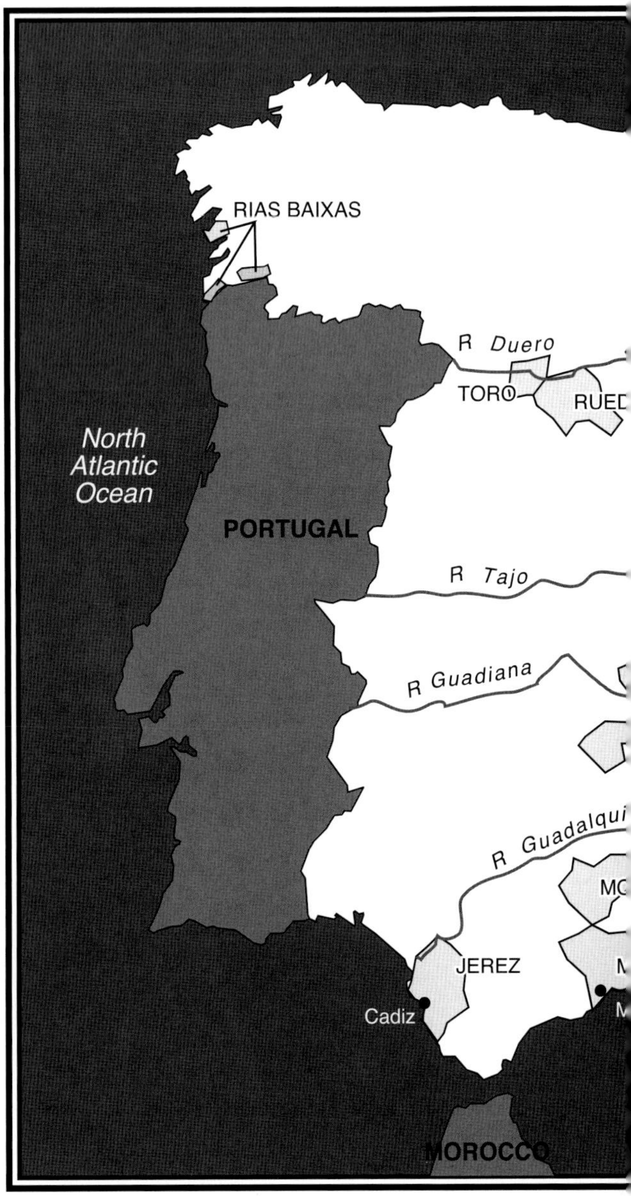

Checking Out Spanish Wines by Region

Besides Rioja, the following regions are an important part of the wine quality picture in Spain today, and their wines are generally available (see figure):

- **Ribera del Duero** *(ree BEAR ah dell DWAIR oh)*, now famous for its high quality red wines, has helped to ignite world interest in Spanish wines.
- **Priorato** *(pree or AH toe)*, mountainous and inaccessible, and one of the world's hot new regions for red wine, is north of the city of Tarragona, in northeast Spain.
- **Penedés** *(pen eh DAIS)* is a large producer of both red and white wines, as well as being famous for its sparkling wines (known as *Cava*).
- The **Rías Baixas** *(REE ahse BYCHE ahse)* region of Galicia *(gah LEETH ee ah)* is gaining acclaim for its exciting white wine, Albariño.
- **Navarra** *(nah VAR rah)*, an area long known for its dry rosé wines, is an increasingly strong red wine region.
- **Toro** is quickly emerging as one of Spain's best red wine regions.
- **Rueda** *(ru AE dah)* is known for well-made, inexpensive white wines.

Spain's wine law

Spain's wine laws, like Italy's, provide for a bi-level QWPSR category: *Denominaciónes de Origen* (DO) and a higher classification, *Denominaciónes de Origen Calificada* (DOC), the latter created in 1991. So far, Rioja and Priorato are the only two regions that have been awarded the DOC (also known as DOCa). Wines that don't qualify as DO fall into the table wine category *Vino de la Tierra* (equivalent to the French *Vins de Pays*).

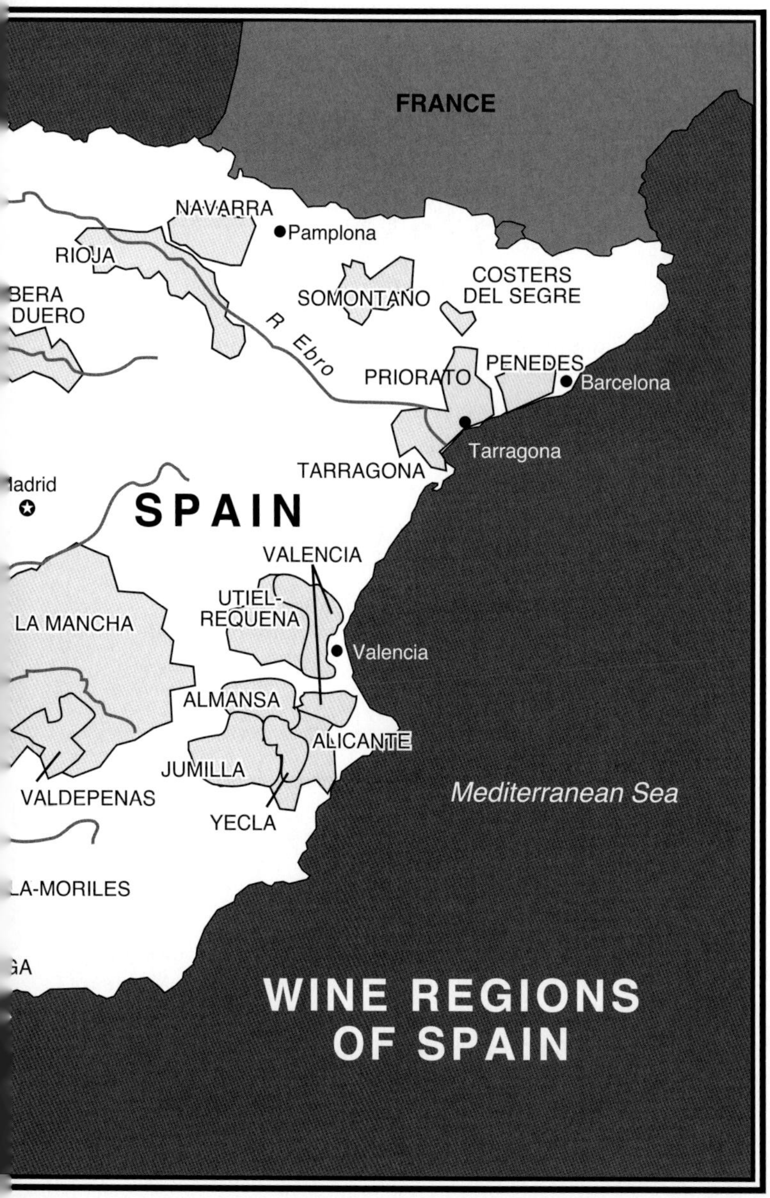

Rioja Rules the Roost

Rioja, in north-central Spain, has historically been the country's major red wine region (even if today Ribera del Duero and Priorato are catching up – fast!). Three-quarters of Rioja's wine is red, 15 per cent *rosado* (rosé), and 10 per cent white.

The principal grape in Rioja is Tempranillo *(tem prah NEE yoh),* Spain's greatest red variety. But regulations permit another three varieties for reds – Garnacha (Grenache), Graciano (Carignan), and Mazuelo – and red Rioja wine is typically a blend of two or more varieties. Regulations aside, some producers now also use Cabernet Sauvignon in their red Rioja.

The Rioja region has three districts: the cooler, Atlantic-influenced Rioja Alavesa and Rioja Alta areas and the warmer Rioja Baja zone. Most of the best Riojas are made from grapes in the two cooler districts, but some Riojas are blended from grapes of all three districts.

Traditional production for red Rioja wine involved many years of aging in small barrels of American oak before release, which created pale, gentle, sometimes tired (but lovely) wines that lacked fruitiness. The trend has been to replace some of the oak aging with bottle aging, resulting in wines that taste much fresher. Another trend, among more progressive winemakers, is to use barrels made of French oak along with barrels of American oak – which has traditionally given Rioja its characteristic vanilla aroma.

Regardless of style, red Rioja wines have several faces according to how long they age before being released from the winery. Some wines receive no oak aging at all and are released young. Some wines age (in oak and in bottle) for two years at the winery and are labeled *crianza;* these wines are still fresh and fruity in style. Other wines age for three years and carry the designation *reserva.* The finest wines age for five years or longer, earning the status of *gran reserva.* These terms appear on the labels – if not on the front label, then on a rear label which is the seal of authenticity for Rioja wines.

TIP

Prices start at around £7 for crianza reds and go up to about £28 for some gran reservas. The best recent vintages for Rioja are 2004, 2001, 1995, 1994, 1989, 1982, and 1981.

The following Rioja producers are particularly consistent in quality for their red wines:

- CVNE (Compañía Vinícola del Norte de España), commonly referred to as CUNE *(COO nay)*
- Bodegas Muga
- R. Lopez de Heredia
- La Rioja Alta
- Marqués de Murrieta Ygay
- Marqués de Riscal

TIP

Most white Riojas these days are merely fresh, neutral, inoffensive wines, but Marqués de Murrieta and R. Lopez de Heredia still make a traditional white Rioja, golden-coloured and oak-aged, from a blend of local white grape varieties, predominantly Viura. Both of these traditional whites are fascinating: flavourful, voluptuous, with attractive traces of oxidation, and capable of aging. They're not everybody's cup of tea, true, but the wines sure have character! They have so much presence that they can accompany foods normally associated with red wine, as well as traditional Spanish food, such as paella or seafood. The Murrieta white sells for about £10, and the Lopez de Heredia is about £12.

Ribera del Duero Challenges

Ribera del Duero, two hours north of Madrid by car, is one of Spain's most dynamic wine regions. Perhaps nowhere else in the world does the Tempranillo grape variety reach such heights, making wines with body, deep colour, and finesse. For many years, one producer, the legendary Vega Sicilia, dominated the Ribera del Duero area. In fact, Spain's single most famous great wine is Vega Sicilia's Unico (Tempranillo, with 20 per cent Cabernet Sauvignon) – an intense, concentrated, tannic red wine with enormous longevity; it ages for ten years in casks and then sometimes ages further in the bottle before it's released. Unico is available mainly in top Spanish restaurants; if you're lucky enough to find it in a retail shop, it can cost about £1880 – a bottle, that is. Even Unico's younger, less intense, and more available sibling, the Vega Sicilia Valbuena, retails for about £60.

TIP

Vega Sicilia is no longer the only renowned red wine in Ribera del Duero. Alejandro Fernández's Pesquera, entirely Tempranillo, has earned high praise over the past 15 years. Pesquera is a big, rich, oaky, tannic wine with intense fruit character. The reserva sells for about £17, while the younger Pesquera is £12. The reserva of Fernández's other winery in the area, Condado de Haza, sells for about £22. Three other fine producers of Ribera del Duero are Bodegas Mauro, Viña Pedrosa, and Bodegas Téofilo Reyes, who all make red wines that rival Pesquera.

Priorato: Emerging from the Past

Back in the twelfth century, monks founded a monastery (or "priory") in the harsh, inaccessible Sierra de Montsant Mountains, about 100 miles southwest of Barcelona in the Catalonia region, and planted vines on the steep hillsides. As time passed, the monastery closed, and the vineyards were abandoned because life was simply too difficult in this area (which in time became known as Priorat, or Priorato).

Cut to the twentieth century – in fact about 25 short years ago. Enterprising winemakers, among them Alvaro Palacios, rediscovered the area and decided that conditions are ideal for making powerful red wines, especially from old vines planted by locals early in the twentieth century.

No Spanish wine region has been in the spotlight lately more than Priorato. And yet Priorato hasn't become a tourist destination, because it's so inaccessible. The region's volcanic soil, composed mainly of slate and schist, is so infertile that not much other than grapes can grow there. The climate is harshly continental: very hot, dry summers and very cold winters. The steep slopes must be terraced; many vineyards can be worked only by hand. And grape yields are very low.

Amazingly rich, powerful red wines – made primarily from Garnacha and Carignan, two of Spain's native varieties – have emerged from this harsh landscape. Many are as rugged as the land, with high tannin and alcohol; some wines are so high in alcohol that they have an almost Port-like sweetness. Because winemaking in Priorato isn't cost-effective, to say the least, and the quantities of each wine are so small, the wines are necessarily quite expensive; prices begin at about £25.

Five Other Spanish Regions to Watch

The action in Spanish wines – especially when value is your concern – definitely doesn't end with Rioja, Ribera del Duero, and Priorato.

Penedés

The Penedés wine region is in Catalonia, south of Barcelona. It's the home of most Spanish sparkling wines, known as *Cava*.

Any discussion of Penedés' still wines must begin with Torres, one of the world's great family-owned wineries. Around 1970, Miguel Torres pioneered the making of wines in Spain from French varieties, such as Cabernet Sauvignon and Chardonnay, along with local grapes, such as Tempranillo and Garnacha.

TIP

Priorato reds to look for include Clos Mogador, Clos Erasmus, Alvaro Palacios, Clos Martinet, l'Hermita, Morlanda, Mas d'En Gil, and Pasanau.

TIP

All the Torres wines are clean, well made, reasonably priced, and widely available. They start in the £6 range for the red Sangre de Toro (Garnacha – Carignan) and Coronas (Tempranil-lo–Cabernet Sauvignon) and the white Viña Sol. The top-of-the-line Mas La Plana Black Label, a powerful yet elegant Cabernet Sauvignon, costs about £30.

TIP

Albariños to look for include Bodega Morgadío, Lusco, Bodegas Martin Codax, Fillaboa, Pazo de Señorans, Pazo San Mauro, Pazo de Barrantes, and Vionta. All are in the £10 to £14 range.

Freixenet, the leading Cava producer, is now also in the still wine business. Its wines include the inexpensive René Barbier brand varietals and two fascinating wines from Segura Viudas (a Cava brand owned by Freixenet), both £9 to £10. Creu de Lavit is a subtle but complex white that's all Xarel-lo (pronounced *sha REL lo*), a native grape used mainly for Cava production. The red Mas d'Aranyo is mainly Tempranillo. We particularly recommend Creu de Lavit.

Rías Baixas: The white wine from Galicia

Galicia, in northwest Spain next to the Atlantic Ocean and Portugal, wasn't a province known for its wine. But from a small area called Rías Baixas *(REE ahse BYCHE ahse)*, tucked away in the southern part of Galicia, an exciting, new white wine has emerged – Albariño, made from the Albariño grape variety. Rías Baixas is, in fact, one of the world's hottest white wine regions. In this case, "hot" means "in demand," not to describe the climate, because Rías Baixas is cool and damp a good part of the year, and verdant year-round.

This region now boasts about 200 wineries, compared to only 60 just a decade ago. Modern winemaking, the cool climate, and low-yielding vines have combined to make Albariño wines a huge success. This lively, (mainly) unoaked white, with its vivid, floral aromas and flavours reminiscent of apricots, white peaches, pears, and green apples, is a perfect match with seafood and fish. The Albariño grape – known as Alvarinho in northern Portugal (south of Rías Baixas) – makes wines that are fairly high in acidity, which makes them fine *apéritif* wines.

Navarra

Once upon a time, the word *Navarra* conjured up images of inexpensive, easy-drinking dry rosé wines (or, to the more adventurous, memories of running the bulls in Pamplona, Navarra's capital city). Today, Navarra, just northeast of Rioja, is known for its red wines, which are similar to, but somewhat less expensive than, the more famous wines of Rioja.

TIP

Many Navarra reds rely on Tempranillo, along with Garnacha, but you can also find Cabernet Sauvignon, Merlot, and various blends of all four varieties in the innovative Navarra region. Look for the wines of the following three Navarra producers: Bodegas Julian Chivite *(HOO lee ahn cha VEE tay)*, Bodegas Guelbenzu *(gwel ben ZOO)*, and Bodegas Magana.

TIP

The Verdejo from Rueda

The Rueda region, west of Ribera del Duero, produces one of Spain's best white wines from the Verdejo grape. The wine is clean and fresh, has good fruit character, and sells for an affordable £5 to £6. The Rioja producer Marquis de Riscal makes one of the leading and most available examples.

El Toro

The Toro region in northwest Spain, west of Ribera del Duero, made wines in the Middle Ages that were quite famous in Spain. But it's a hot, arid area with poor soil, so winemaking was practically abandoned there for centuries. In Spain's current wine boom, Toro has been rediscovered. Winemakers have determined that the climate and soil are actually ideal for making powerful, tannic red wines — mainly from the Tempranillo variety — which rival the wines of Toro's neighbours in Ribera del Duero. Toro producers to buy include Bodegas Fariña, Vega Sauco, Estancia Piedra, Bodegas y Viñas Dos Victorias, Gil Luna, and Dehesa La Granja (owned by Pesquera's Alejandro Fernandez).

Decoding Spanish wine labels

You see some of the following terms on a Spanish wine label:

Blanco: White

Bodega: Winery

Cosecha *(coh SAY cha)* or **Vendimia** *(ven DEE me yah):* The vintage year

Crianza *(cree AHN zah):* For red wines, this means that the wine has aged for two years with at least six months in oak; for white and rosé wines, crianza means that the wines aged for a year with at least six months in oak. (Some regions have stricter standards.)

Gran reserva: Wines produced only in exceptional vintages; red wines must age at least five years, including a minimum of two years in oak; white gran reservas must age at least four years before release, including six months in oak.

Reserva: Wines produced in the better vintages; red reservas must age a minimum of three years, including one year in oak; white reservas must age for two years, including six months in oak.

Tinto (TEEN toe): Red

Wines in Germany and Portugal

IN THIS ARTICLE

- *Taking a look at Germany's unique ways*
- *Grabbing great finds from Portugal*

In the past, no one ever used the phrase *European wine* when talking generally about the wines of France, Italy, Spain, Portugal, and Germany. The wines had nothing in common. But today, Europe has unified, and the wines of the European Union member countries now share a common legislative umbrella.

When you compare Europe's wines to non-European, or *New World,* wines, the diverse wines of Europe have many things in common after all. Most European wines are usually named for their place of production instead of their grape; European winemaking is tethered to tradition and regulations; the wines, for the most part, have local flavour rather than conforming to an international concept of how wine should taste; and these wines are relatively low in fruitiness. European wines tend to embody the traditions of the people who make them and the flavours of the place where their vines grow, unlike New World wines, which tend to embody grape variety and a general fruitiness of flavour.

Despite these similarities among European wines, the countries of Europe each make distinctly different wines. In this chapter, the wines of Germany and Portugal take centre stage.

Germany: Europe's Individualist

German wines march to the beat of a different drummer. They come in mainly one colour: white. They're fruity in style, low in alcohol, rarely oaked, and often off-dry or sweet. Their labels carry grape names, which is an anomaly in Europe.

Germany is the northernmost major wine-producing country in Europe – which means that its climate is cool. Except in warmer pockets of Germany, red grapes don't ripen adequately, which is the reason most German wines are white. The climate is also erratic from year to year, meaning that vintages do matter for fine German wines. Germany's finest vineyards are situated along rivers such as the Rhine and the Mosel, and on steep, sunny slopes, to temper the extremes of the weather and help the grapes ripen.

Riesling and its cohorts

In Germany's cool climate, the noble Riesling *(REESE ling)* grape finds true happiness. Riesling represents little more than 20 per cent of Germany's vineyard plantings.

Another major, but less distinguished, German variety is Müller-Thurgau (pronounced *MOOL lair TOOR gow*), a crossing between the Riesling and Silvaner (or possibly Chasselas) grapes. Its wines are softer than Riesling's with less character and little potential for greatness.

After Müller-Thurgau and Riesling, the most-planted grapes in Germany are Silvaner, Kerner, Scheurebe *(SHOY reb beh),* and Ruländer (Pinot Gris). Among Germany's red grapes, Spätburgunder (Pinot Noir) is the most widely planted, mainly in the warmer parts of the country.

Germany's wine laws and wine styles

Germany's classification system isn't based on the French AOC system, as those of most other European countries are. Like most European wines, German wines are named after the places they come from – in the best wines, usually a combination of a village name and a vineyard name, such as Piesporter (town) Goldtröpfchen (vineyard).

Unlike most European wines, however, the grape name is also usually part of the wine name (as in Piesporter Goldtröpfchen *Riesling*). And the finest German wines have yet another element in their name – a *Prädikat (PRAY di cat),* which is an indication of the ripeness of the grapes at harvest (as in Piesporter Goldtröpfchen Riesling *Spätlese*). Wines with a Prädikat hold the highest rank in the German wine system.

Germany's system of assigning the highest rank to the ripest grapes is completely different from the concept behind most other European systems, which is to bestow the highest status on the best vineyards or districts. Germany's system underscores the country's grape-growing priority: Ripeness – never guaranteed in a cool climate – is the highest goal.

REMEMBER

German wine law divides wines with a Prädikat into six levels. At the three highest Prädikat levels, the amount of sugar in the grapes is so high that the wines are inevitably sweet. Many people, therefore, mistakenly believe that the Prädikat level of a German wine is an indication of the wine's sweetness. In fact, the Prädikat is an indication of the amount of sugar in the *grapes at harvest,* not the amount of sugar in the wine. At lower Prädikat levels, the sugar in the grapes can ferment fully, to dryness, and for those wines there is no direct correlation between Prädikat level and sweetness of the wine.

Wines whose (grape) ripeness earns them a Prädikat are categorized as QmP wines *(Qualitätswein mit Prädikat),* translated as *quality wines with special attributes* (their ripeness). They are QWPSR wines in the eyes of the EU. When the ripeness of the grapes in a particular vineyard isn't sufficient to earn the wine a Prädikat name, the wine can still qualify as a "quality wine" in Germany's second QWPSR tier, called QbA *(Qualitätswein bestimmter Anbaugebiet),* translated as *quality wine from a special region.* (Refer to the section, "Germany's wine regions," for the names of the main regions.) Often just the term *Qualitätswein* appears on labels of QbA wines, without the words *bestimmter Anbaugebiet* — and the name of the region will always appear.

Less than 10 per cent of Germany's wine production falls into the lower, table wine categories *Landwein* (table wines with geographic indication) or *Deutscher Tafelwein.* Most of the inexpensive German wines that you see in wine shops are QbA wines.

Dry, half-dry, or gentle

The common perception of German wines is that they're all sweet. Yet many German wines taste dry, or fairly dry. In fact, you can find German wines at just about any sweetness or dryness level you like.

Most inexpensive German wines, such as Liebfraumilch, are light-bodied, fruity wines with pleasant sweetness – wines that are easy to enjoy without food. The German term for this style of wine is *lieblich,* which translates as "gentle" – a poetic but apt descriptor. The very driest German wines are called *trocken* (dry). Wines that are sweeter than trocken but dryer than lieblich are called *halbtrocken* (half-dry). The words *trocken* and *halbtrocken* sometimes appear on the label, but not always.

A bit of sweetness in German wines can be appealing – and in fact can improve the quality of the wine. That's because sweetness undercuts the wines' natural high acidity and gives the wines better balance. In truth, most off-dry German wines don't really taste as sweet as they are, thanks to their acidity.

TIP

You can make a good stab at determining how sweet a German wine is by reading the alcohol level on the label. If the alcohol is low – about 9 percent, or less – the wine probably contains grape sugar that didn't ferment into alcohol and is therefore sweet. Higher alcohol levels suggest that the grapes fermented completely, to dryness.

What's noble about rot?

Wine connoisseurs all over the world recognize Germany's sweet, dessert-style wines as among the greatest wines on the face of the earth. Most of these legendary wines owe their sweetness to an ugly but magical fungus known as *botrytis cinerea,* pronounced *bo TRY tis sin eh RAY ah,* commonly called *noble rot.*

Noble rot infects ripe grapes in late autumn if a certain combination of humidity and sun is present. This fungus dehydrates the berries and concentrates their sugar and their flavours. The wine from these infected berries is sweet, amazingly rich, and complex beyond description. It can also be expensive: £70 a bottle or more.

Another way that Nature can contribute exotic sweetness to German wines is by freezing the grapes on the vine in early winter. When the frozen grapes are harvested and pressed, most of the water in the berries separates out as ice. The sweet, concentrated juice that's left to ferment makes a luscious sweet Prädikat-level wine called *Eiswein* (literally, ice wine). Eisweins differ from BAs and TBAs because they lack a certain flavour that derives from botrytis, sometimes described as a honeyed character.

Both botrytised wines and Eisweins are referred to as *late-harvest wines,* not only in Germany but all over the world, because the special character of these wines comes from conditions that normally occur only when the grapes are left on the vine beyond the usual point of harvest.

Germany's wine regions

Germany has 13 wine regions – 11 in the west and 2 in the eastern part of the country (see figure).

The most famous of these 13 are the Mosel-Saar-Ruwer region, named for the Mosel River and two of its tributaries, along which the region's vineyards lie; and the Rheingau region, along the Rhine River. The Rhine River lends its name to three other German wine regions, Rheinhessen, the Pfalz (formerly called the Rheinpfalz), and the tiny Mittelrhein region.

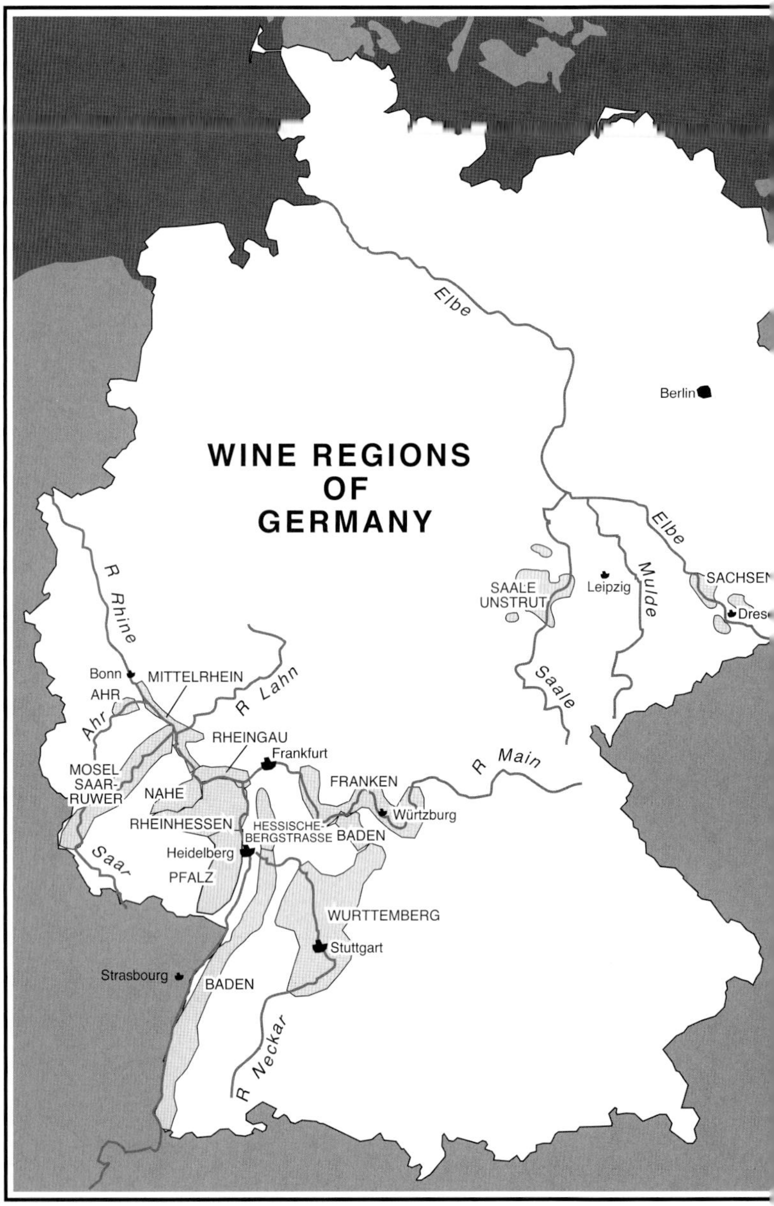

- **Mosel-Saar-Ruwer:** The Mosel-Saar-Ruwer *(MO zel zar ROO ver)* is a dramatically beautiful region, its vineyards rising steeply on the slopes of the twisting and turning Mosel River. The wines of the region are among the lightest in Germany (usually containing less than 10 per cent alcohol); they're generally delicate, fresh, and charming. Riesling dominates the Mosel-Saar-Ruwer with 57 per cent of the plantings. Wines from this region are instantly recognizable because they come in green bottles rather than the brown bottles that other German regions use.
- **Rheingau:** The Rheingau *(RYNE gow)* is among Germany's smaller wine regions. It, too, has some dramatically steep vineyards bordering a river, but here the river is Germany's greatest wine river, the Rhine. The Riesling grape occupies more than 80 per cent of the Rheingau's vineyards, many of which are south-facing slopes that give the Riesling grapes an extra edge of ripeness. Rheingau wine styles tend toward two extremes: trocken wines on the one hand and sweet late-harvest wines on the other. Recommended Rheingau producers include Georg Breuer, Knyphausen, Franz Küntsler, Schloss Schönborn, Leitz, and Robert Weil.

TIP

The Mosel boasts dozens of excellent winemakers who produce really exciting Riesling wines. Some favourites, in alphabetical order, include Egon Müller, Meulenhof, Dr. Fischer, J.J. Prüm, Friedrich Wilhelm Gymnasium, Reichsgraf Von Kesselstatt, Karlsmühle, Willi Schaefer, Dr. Loosen, Selbach-Oster, Maximin Grünhauser, Zilliken, and Merkelbach.

A secret code of German place-names

If you don't speak German and you don't know German geography intimately, deciphering German wine names is tricky, to say the least. But here's a bit of information that can help. In the German language, the possessive is formed by adding the suffix -er to a noun. When you see names like Zeller or Hochheimer – names that end in -er – on a wine label, the next word is usually a vineyard area that "belongs" to the commune or district with the -er on its name (Zell's Swartze Katz, Hochheim's Kirchenstück). The name of the region itself always appears on labels of QbA and Prädikat wines.

- **Rheinhessen:** Rheinhessen *(RYNE hess ehn)* is Germany's largest wine region, producing huge quantities of simple wines for everyday enjoyment. Liebfraumilch originated here, and it's still one of the most important wines of the region, commercially speaking. The Rheinhessen's highest quality wines come from the Rheinterrasse, a vineyard area along the river. Producers from that area who are particularly good include Gunderloch, Heyl Zu Herrnsheim, and Strub.

- **Pfalz:** Almost as big as the Rheinhessen, the Pfalz *(fallz)* has earned somewhat more respect from wine lovers for its fairly rich and full-bodied white wines and its very good reds – all of which owe their style to the region's relatively warm climate. Müller-Thurgau, Riesling, Silvaner, and Kerner are among the most planted grape varieties of the Pfalz, but qualitatively Scheurebe and Blauburgunder (Pinot Noir) are important. To experience the best of the Pfalz, look for wines from Dr. Bürklin-Wolf, Rainer Lingenfelder, Müller-Catoir, and Basserman-Jordan.

- **Nahe:** One other German region of importance for the quality of its wines is Nahe (NAH heh), named for the Nahe River and situated west of Rheinhessen. The Riesling wines here are relatively full and intense. Favorite producers include Diel, Kruger-Rumpf, Prinz zu Salm-Dahlberg, and Dönnhoff.

Portugal: More Than Just Port

Portugal is justifiably famous for its great dessert wine, Port. But gradually, wine lovers are discovering the other dimensions of Portuguese wine — its dry wines, especially its reds. Most of these wines come from native Portuguese grape varieties, of which the country has hundreds. Portugal's well-priced wines will probably play a larger role in world wine markets in the twenty-first century.

Portugal's highest rank for wines is *Denominação de Origen* (DO), which has been awarded to the wines of 32 regions. The table wine category includes eight *Vinho Regional* (VR) regions, equivalent to France's *Vin de Pays,* and the simple *Vinho de Mesa* (table wines).

Portugal's "green" white

The Minho region, Vinho Verde's home, is in the northwest corner of Portugal, directly across the border from the Rías Baixas wine region of Spain. (The region is particularly verdant because of the rain from the Atlantic Ocean – one theory behind the wine's name.)

TIP

On hot summer evenings, the most appropriate wine can be a bottle of bracing, slightly effervescent, white Vinho Verde *(VEEN yo VAIRD)*. The high acidity of Vinho Verde refreshes your mouth and particularly complements grilled fish or seafood.

Two styles of white Vinho Verde exist on the market. The most commonly found brands (Aveleda and Casal Garcia), which sell for £4 or £5, are medium-dry wines of average quality that are best served cold.

The more expensive Vinho Verdes (£9 to £12) are varietal wines made from either the Alvarinho grape (Rías Baixas's Albariño), Loureiro, or Trajadura. They're more complex, dryer, and more concentrated than basic Vinho Verde, and are Portugal's best whites. Unfortunately, these finer wines are more difficult to find than the inexpensive ones; look for them in better wine shops or in Portuguese neighborhoods — or on your next trip to Portugal!

WARNING!

The majority of wines from Vinho Verde are red. However, these wines are *highly* acidic ; you definitely need to acquire a taste for them.

Noteworthy Portuguese red wines

Possibly the best dry red wine in Portugal, Barca Velha, comes from the Douro region, where the grapes for Port (officially known as *Porto*) grow. Made by the Ferreira Port house, Barca Velha is a full-bodied, intense, concentrated wine that needs years to age – Portugal's version of Vega Sicilia's Unico, but at a considerably lower price £40 to £44). Like Unico, not much is made, and it's produced only in the best vintages.

The Douro region boasts other terrific dry red wines, most of them fairly new and based on grapes traditionally used for Port. Brands to look for include Quinta do Vale D. Maria, Quinta do Vallado, Quinta do Crasto, Quinta do Cotto, Quinta de la Rosa, Quinta do Vale Meão, Quinta de Roriz, Quinta da Leda Vale do Bomfim and Chryseia.

TIP

Fortunately, the Port house of Ramos Pinto (now owned by Roederer Champagne) makes inexpensive, top-quality, dry red Douro wines that are readily available. Duas Quintas (about £7) has ripe, plummy flavours and a velvety texture; it's surprisingly rich but supple, and it's a great value.

TIP

Other good red Portuguese wines to try include

- **Quinta do Carmo:** The majority owner of this estate in the dynamic Alentejo region in southern Portugal is Château Lafite-Rothschild. A rich, full-bodied wine, it sells for £15. Don Martinho, a second-label wine from the estate, is less than half the price of Quinta do Carmo.
- **Quinta de Pancas:** One of the few Cabernet Sauvignons in Portugal, Quinta de Pancas comes from the Alenquer region, north of Lisbon; it sells for about £9.
- **Quinta de Parrotes:** Made from the local Castelão Frances grape variety, Quinta de Parrotes, from the same estate in Alenquer as the Quinta de Pancas, is a steal at £6.
- **Quinta da Bacalhôa:** An estate-bottled Cabernet Sauvignon-Merlot from the esteemed Portuguese winemaker Joào Pires in Azeitao (south of Lisbon), Bacalhôa has the elegance of a Bordeaux; it sells for £17.
- **The red wines of J.M. da Fonseca Successores** (no relation to the Fonseca Port house): This firm is producing some of the best red wines in Portugal. Look for Quinta da Camarate, Morgado do Reguengo, Tinto Velho Rosado Fernandes, and all da Fonseca's Garrafeiras.
- **The wines of Joao Portugal Ramos:** A tireless winemaker who consults for various wineries and also owns three properties, Ramos has a golden touch and yet maintains the typicity of his wines. Some wines sell under his own name; others are Marquês de Borba and Vila Santa.

Wines Elsewhere in Europe

IN THIS ARTICLE

- *Checking out Switzerland's wines*
- *Enjoying Hungary's wine*
- *Drinking Austria's finest*
- *Living it up with Greek wines*

France, Italy, and Spain are the big wine producers. But that doesn't mean that wine production is limited to those countries. Other countries, such as Switzerland, Austria, Hungary, and Greece, also produce wines worthy of note as well.

In case you're wondering, France, Italy, and Spain are each covered in their own articles, while Portugal and Germany share space in another article.

Switzerland's Stay-at-Home Wines

Nestled between Germany, France, and Italy, Switzerland is in a perfectly logical location for growing grapes and making fine wine. Vineyards grace the country's three faces – French-speaking, German-speaking, and Italian-speaking. But few wine lovers outside of Switzerland have much opportunity to taste Swiss wines because the production is tiny and because the wines are so popular within Switzerland itself.

About half of Switzerland's wines are white; most are made from Chasselas – a grape cultivated with much less distinction in Germany, eastern France, and the Loire Valley. In Switzerland, Chasselas wines tend to be dry, fairly full-bodied, and unoaked, with mineral and earthy flavours. Other white grapes include Pinot Gris, Sylvaner, Marsanne, Petit Arvine, and Amigne – the latter two indigenous to Switzerland. Merlot is an important red grape (especially in the Italian-speaking Ticino region), along with Pinot Noir and Gamay.

Because of Switzerland's varied terrain (hills of varying altitudes, large lakes, sheltered valleys), numerous microclimates exist. Wine styles therefore vary, from relatively full-bodied reds and whites to delicate, crisp white wines.

Switzerland's major wine regions include the Vaud, along Lake Geneva; Valais, to the east, along the Rhône River; Neuchâtel, in western Switzerland, north of the Vaud; Ticino, in the south, bordering Italy; and Thurgau in the north, bordering Germany.

When you do find a bottle of Swiss wine, you may be surprised to discover how costly it is, reflecting high production costs. (But quality is generally also high.)

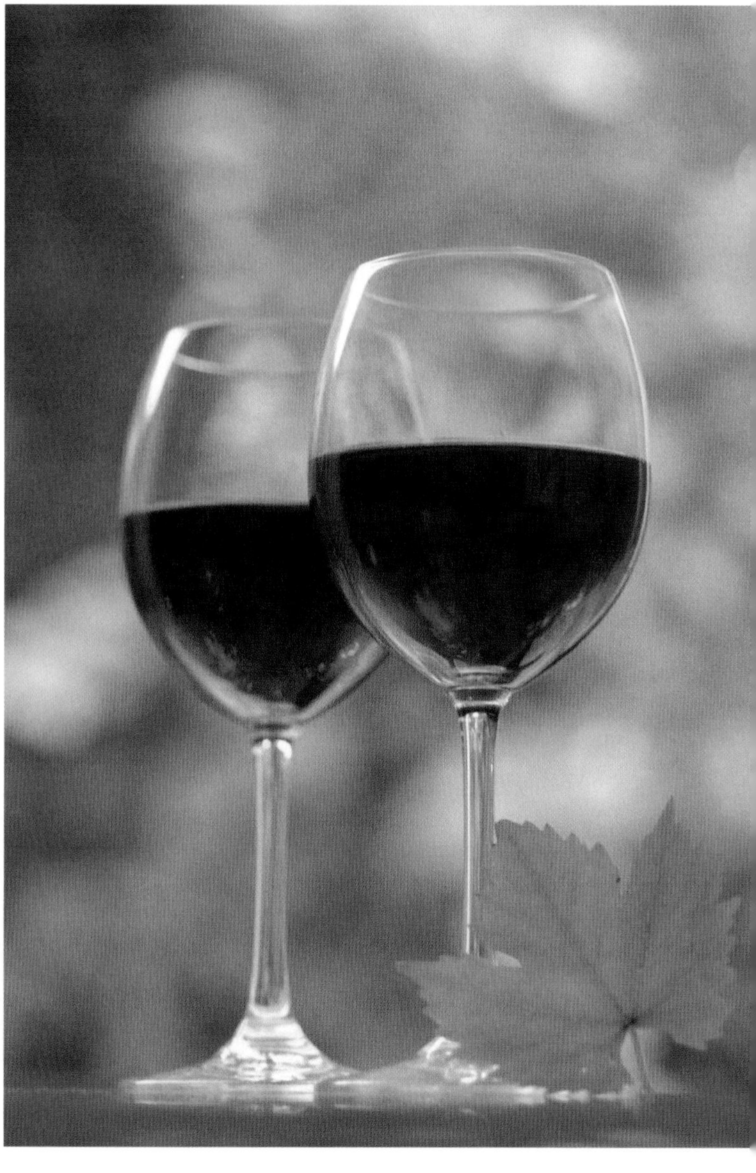

Austria's Exciting Whites (and Reds)

Austria is an exciting wine country with wonderful wines, as well as gorgeous vineyard regions, warm people, and the classic beauty of Vienna. What makes Austrian wines all the more interesting is how they're evolving, as winemakers gradually discover how to best express their land and their grapes through wine.

Austria makes less than 1 per cent of all the wine in the world — about 28 million cases a year. All the wine comes from the eastern part of the country, where the Alps recede into hills, and most of it comes from small wineries. Although some inexpensive Austrian wines do make their way to export markets, the Austrians have embraced a high-quality image, and most of their wines therefore command premium prices.

While the excellence of Austria's sweet whites has long been recognized, its dry whites and reds have gained recognition only in the past two decades. Reds are in the minority, claiming about 25 per cent of the country's production, because many of Austria's wine regions are too cool for growing red grapes. Red wines hail mainly from the area of Burgenland, bordering Hungary, one of the warmest parts of the country. They're medium- to full-bodied, often engagingly spicy, with vivid fruity flavour – and often the international touch of oaky character. Many of them are based on unusual, native grape varieties such as the spicy Blaufrankish (Lemberger), the gentler St. Laurent, or Blauer Zweigelt (a crossing of the other two).

Austria's white wines — apart from the luscious, late-harvest dessert wines made from either botrytised, extremely ripe, or dried grapes — are dry wines ranging from light- to full-bodied that are generally unoaked.

The country's single most important grape variety is the native white Grüner Veltliner. Its wines are full-bodied yet crisp, with rich texture and herbal or sometimes spicy-vegetal flavours (especially green pepper). They're extremely food-friendly, and usually high quality. Some people in the wine trade have nicknamed Grüner Veltliner "GruVe."

Riesling, grown mainly in the region of Lower Austria, in the northeast, is another key grape for quality whites. In fact, some experts believe that Austria's finest wines are its Rieslings (while others prefer Grüner Veltliner). Other grape names that you may see on bottles of Austrian wine include Müller-Thurgau, which makes characterful dry whites; Welschriesling, a grape popular in Eastern Europe for inexpensive wines that achieves high quality only in Austria; Pinot Blanc, which can excel here; and Muscat. Sauvignon Blanc is a specialty of the region of Styria, in the south, bordering Slovenia.

In some parts of Austria, for example in the Wachau district, along the Danube River, wines are named in the German system – a town name ending in *-er* followed by a vineyard name and a grape variety. In other parts of Austria, the wine names are generally a grape name (or, increasingly, a proprietary name) followed by the name of the region.

The Re-Emergence of Hungary

Of all the wine-producing countries in Eastern Europe that broke free from Communism in the late 1980s and early 1990s and have resumed wine production under private winery ownership, Hungary seems to have the greatest potential. In addition to a winemaking tradition that dates back to pre-Roman times, Hungary has a wealth of native and international grape varieties and plenty of land suited to vineyards, with a wide range of climates, soils, and altitudes.

The Hungarians are a proud and creative people. Their wine consumption has increased significantly since the country gained independence, fueling an improvement in wine quality. International investment in vineyards and wineries has also made a huge contribution.

Hungary produces the equivalent of about 68 million cases of wine a year, most of which is white. Although the country is northerly its climate is relatively warm because the country is landlocked and nearly surrounded by mountains. Three large bodies of water do affect the microclimate of certain wine regions: Lake Neusiedel, between Hungary and Austria in the northwest; Lake Balaton, Europe's largest lake, in the center of Hungary's western half (which is called Transdanubia); and the Danube River, which runs north to south right through the middle of the country. Hungary has 22 official wine regions, but their names are not yet particularly important outside Hungary.

The one Hungarian wine region that does have international fame is Tokaj-Hegyalja *(toe KYE heh JAH yah),* which takes its name from the town of Tokaj and owes its reputation to its world-class dessert wine, Tokaji Azsu *(toe KYE as ZOO).* The word *Aszu* refers to botrytised grapes. The wine comes from Furmint and Harslevelu grapes, both native white varieties, and sometimes Muscat grapes, that have been infected by botrytis. This region also makes dry table wines, such as the varietal Tokaji Furmint.

Tokaji Azsu wines are labeled as three, four, five, or six Puttonyos, according to their sweetness, with six Puttonyos wines being the sweetest. (*Puttonyos* are baskets used to harvest the botrytised grapes, as well as a measure of sweetness.) All Tokaji Azsu wines sell in 500 ml bottles, and they range in price from about £22 to £94 per bottle, depending on their sweetness level.

Tokaji Azsu wines vary not only according to their sweetness, but also according to their style. Some wines have fresher, more vibrant fruity character, for example; some have aromas and flavours that suggest dried fruits; some have the smoky character and tannin of new oak barrels; and some have complex non-fruity notes such as tea leaves or chocolate. This range of styles is due mainly to different winemaking techniques among producers.

Beyond the famous Tokaj-Hegyalja region, Hungary has numerous other wine regions that produce a range of dry and semi-dry wines, both white and red. Most of these wines are named for their grape variety and are quite inexpensive. Kadarka is Hungary's best-known native red grape variety.

The Glory That Is Greece

The country that practically invented wine, way back in the seventh century BC, is now an emerging wine region today. But that's the way it is. Greece never stopped making wine for all those centuries, but its wine industry took the slow track, inhibited by Turkish rule, political turmoil, and other real-life issues. The modern era of Greek winemaking began only in the 1960s, and it has made particularly strong strides in the past decade. Today, Greek wines are worth knowing.

Although Greece is a southern country and famous for its sunshine, its grape-growing climate is actually quite varied, because many vineyards are situated at high altitudes where the weather is cooler. (Most of Greece is mountainous, in fact.) Its wines are mainly (60 per cent) white; some of those whites are sweet dessert wines, but most are dry.

One of Greece's greatest wine assets – and handicaps, at the same time – is its abundance of native grape varieties, over 300 of them. Only Italy has more indigenous grape varieties. These native grapes make Greek wines particularly exciting for curious wine lovers to explore, but their unfamiliar names make the wines difficult to sell. Fortunately for the marketers, Greece also produces wines from internationally-famous grape varieties such as Chardonnay, Merlot, Syrah, and Cabernet Sauvignon, and those wines can be very good. These days, however, producers seem more committed than ever to their native varieties rather than to international grapes.

Of Greece's many indigenous grape varieties, four in particular stand out as the most important — two white and two red varieties:

- **Assyrtiko** (*ah SEER tee koe*): A white variety that makes delicate, bone dry, crisp, very long-lived wines with citrusy and minerally aromas and flavours. Although Assyrtiko grows in various parts of Greece, the best Assyrtiko wines come from the volcanic island of Santorini. Any wine called Santorini is made at least 90 per cent from Assyrtiko.

- **Moschofilero** (*mos cho FEEL eh roe*): A very aromatic, pink-skinned variety that makes both dry white and pale-coloured dry rosé wines grows mainly around Mantinia, in the central, mountainous Peloponnese region. If a wine is named Mantinia, it must be at least 85 per cent Moschofilero. Wines made from Moschofilero have high acidity and are fairly low in alcohol, with aromas and flavours of apricots and/or peaches. Because they're so easy to drink, Moschofilero wines are a great introduction to Greek wines.

- **Agiorghitiko** (*eye your YEE tee koe*): The name of this grape translates in English to "St. George," and a few winemakers call it that on the labels of wines destined for English-speaking countries. Greece's most-planted and probably most important native red variety, it grows throughout the mainland. Its home turf, where it really excels, is in the Nemea district of the Peloponnese region; any wine named Nemea is entirely from Agiorghitiko. Wines from this variety are medium to deep in colour, have complex aromas and flavours of plums and/or blackcurrants, and often have a resemblance to Cabernet Franc or spicy Merlot wines. Agiorghitiko also blends well with other indigenous or international varieties.

- **Xinomavro** (*ksee NO mav roe*): The most important red variety in the Macedonia region of Northern Greece. Xinomavro produces highly tannic wines with considerable acidity that have been compared to Nebbiolo wines of Piedmont, Italy. Wines made from Xinomavro have complex, spicy aromas, often suggesting dried tomatoes, olives, and/or berries. Xinomavro wines are dark in colour but lighten with age, and have great longevity. Their home base is the Naoussa district of Macedonia; any wine named Naoussa is entirely from Xinomavro.

Other important white indigenous varieties in Greece include Roditis (actually a pink-skinned grape), which makes Patras white; and Savatiano, the most widely planted white grape. Retsina, a traditional Greek wine made by adding pine resin to fermenting grape juice (resulting in a flavour not unlike some oaky Chardonnays), is made mainly from Savatiano. Mavrodaphne is an indigenous Greek red variety that is becoming increasingly important, both for dry and sweet red wines.

Some of the wine regions of Greece whose names you're likely to see on wine labels include

- **Macedonia:** The northernmost part of Greece, with mountainous terrain and cool climates. Naoussa wine comes from here.
- **The Peloponnese:** A large, mainly mountainous, peninsula in southwestern Greece with varied climate and soil. Noteworthy wines include the soft, red Nemea; the dry whites Patras and Mantinia; and the sweet wines Mavrodaphne de Patras (red) and Muscat de Patras (white).
- **Crete:** The largest Greek island, which makes both white and red wines, many of which are varietally-named along with the place-name of Crete.
- **Other Greek Islands:** Besides Crete, the four most important islands that make wine are Santorini, Rhodes, Samos, and Cephalonia.

Many Greek wines today are top-quality, especially the wines of small, independent wineries. Some favourite Greek wine producers (listed alphabetically within regions) include

- From Macedonia: Alpha Estate, Domaine Gerovassilou, Kir Yianni Estate, and Tsantali-Mount Athos Vineyards
- From the Peloponnese: Antonopoulos Vineyards, Gaia Estate (pronounced *YEA ah,* has wineries also in Santorini), Katogi & Strofilia (with operations also in Macedonia), Mercouri Estate, Papantonis Winery, Domaine Skouras, Domaine Spiropoulos, and Domaine Tselepos
- From the islands: Boutari Estates (six estates throughout Greece, including Crete and Santorini), Gentilini (in Cephalonia), and Domaine Sigalas (Santorini)

Table wines with a geographic name are called *vins de pays* (regional wines). Many of Greece's better wines in fact carry a *vins de pays* appellation. Other terms that have formal definitions under Greek wine regulations include *reserve* (QWPSR wines with a minimum two or three years aging, for whites and reds respectively), *grande reserve* (one additional year of aging), and *cava* (a table wine – in the EU sense of being at the lower appellation tier – with the same aging requirements as reserve).

North American Wines

IN THIS ARTICLE

- *Living it up in Napa*
- *Relaxing in Sonoma*

U.S. wines have elevated grape varieties to star status. Until California began naming wines after grapes, Chardonnay, Merlot, Pinot Noir, and Cabernet Sauvignon were just behind-the-scenes ingredients of wine – but now they *are* the wine. Lest anyone think that all wines from a particular grape are the same, however, winemakers have emerged as celebrities who put their personal spin on the best wines. In the California scenario especially, the land – the terroir – has been secondary, at least until recently.

In this article, American wines, particularly California's special wine regions, take centre stage. Oregon, Washington State, New York, and Canada also play a supporting role.

California, USA

When most wine drinkers think about American wine, they think of California. That's not surprising – the wines of California make up about 88 per cent of U.S. wine production.

In sunny California, there's no lack of warm climate for growing grapes. For fine wine production, the challenge is to find areas cool enough, with poor enough soil, so that grapes don't ripen too quickly, too easily, without full flavour development. Nearness to the Pacific Coast and higher altitudes both assure cooler climates more so than latitude does. Fine wines therefore come from vineyards up and down almost the whole length of the state.

The most important fine wine areas and districts include the following (see figure):

- **North Coast:** Napa Valley, Sonoma County, and Mendocino and Lake Counties
- **North-Central Coast:** Livermore and Santa Clara Valleys (San Francisco Bay area), Santa Cruz Mountains, and Monterey County
- **Sierra Foothills**
- **South-Central Coast:** San Luis Obispo County and Santa Barbara County

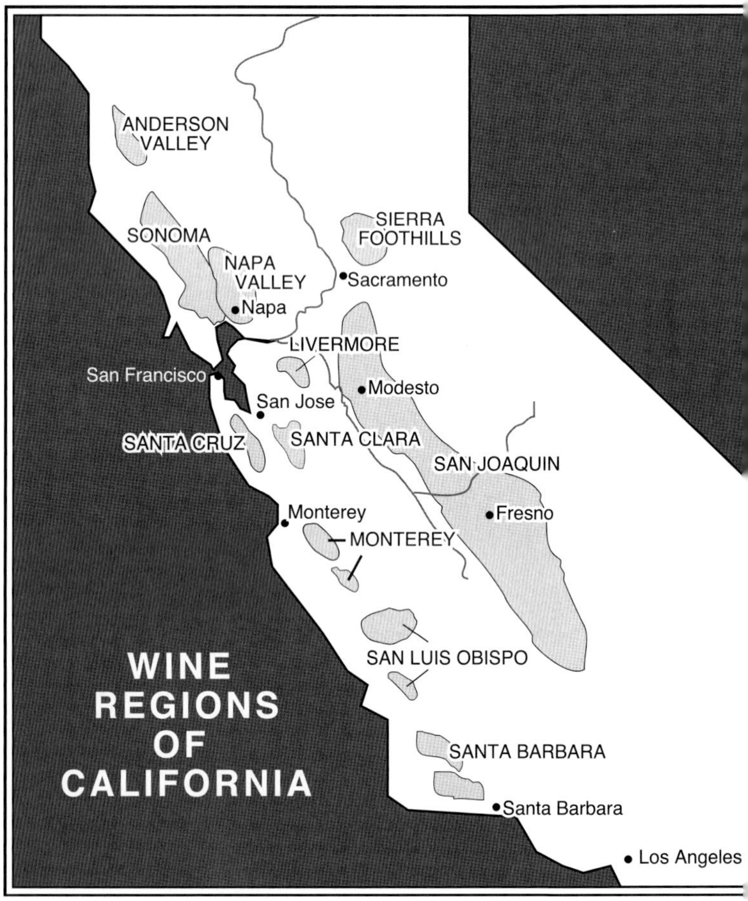

The wine regions of California.

Weather variations from year to year are far less dramatic in California than they are in most European wine regions. One major reason is that rain doesn't fall during the growing season in much of California. (Rain at the wrong time is the usual cause of Europe's poorer vintages.) Using irrigation, winemakers, in effect, control the water to the vines. Ironically, one factor that can cause vintage variation in California is lack of water for irrigation due to drought.

Napa Valley: As tiny as it is famous

Napa Valley is about a 90-minute drive north-east of the beautiful bay city of San Francisco. Many of California's most prestigious wineries – and certainly its most expensive vineyard land – are in the small Napa Valley, where about 240 wineries have managed to find space. (In 1960, Napa Valley had only 25 wineries.) The region's size is actually much tinier than its reputation: Napa produces less than 5 per cent of California's wine grapes.

The southern part of the Valley, especially the Carneros district, is the coolest area, thanks to ocean breezes and mists from the San Pablo Bay. Carneros – which extends westward into Sonoma County – has become the vineyard area of choice for grape varieties that enjoy the cool climate: Chardonnay, Pinot Noir, Merlot, and grapes for sparkling wines. North towards Calistoga – away from the bay influence – the climate gets quite hot (but always with cool nights).

Wineries and vineyards occupy almost every part of Napa Valley. Many vineyards are on the valley floor, some are in the hills and mountains to the west (the Mayacamas Mountains), and some are in the mountains to the east (especially Howell Mountain).

Almost everyone in Napa who makes table wine makes a Cabernet Sauvignon and a Chardonnay, and many Napa producers now also make Merlot.

The six most important wines in Napa are the two whites, Chardonnay and Sauvignon Blanc (often labeled Fumé Blanc), and the four red wines, Cabernet Sauvignon, Merlot, Pinot Noir (mainly from cool Carneros), and Zinfandel. But blended wines have become increasingly important in the last 15 years. If red, these blends are usually made from red Bordeaux varieties (Cabernet Sauvignon, Cabernet Franc, Merlot, and sometimes even Malbec and Petit Verdot). If white, they're usually made from the white Bordeaux grapes (Sauvignon Blanc and Sémillon). Some of these blends are referred to as Meritage wines – not just in Napa but across the United States – although few carry the word *Meritage* on their labels.

Down-to-earth in Sonoma

If you leave San Francisco over the beautiful Golden Gate Bridge, you'll be in Sonoma in an hour. The differences between Napa and Sonoma are remarkable. Many of Napa's wineries are showy (even downright luxurious), but most of Sonoma's are rustic, country-like, and laid-back. The millionaires bought into Napa; Sonoma is just folks (with some exceptions, of course).

On the other hand, the famously successful Gallo is also in Sonoma, and so are Sebastiani, Glen Ellen, Korbel, Kendall-Jackson, Simi, and Jordan wineries – not exactly small time operations!

Sonoma is more than twice as large as Napa, it's more spread out, and it has almost as many wineries — more than 200. Its climate is similar to Napa's, except that some areas near the coast are definitely cooler. Although there's plenty of Chardonnay, Cabernet Sauvignon, and Merlot in Sonoma, the region's varied microclimates and terrain have allowed three other varieties – Pinot Noir, Zinfandel, and Sauvignon Blanc — to excel.

Mendocino and Lake Counties

Lake County, dominated by Clear Lake, is Napa's neighbor to the north, and Mendocino County is directly north of Sonoma.

The cool Anderson Valley in Mendocino County is ideal for growing Chardonnay, Pinot Noir, Gewürztraminer, and Riesling, and for the production of sparkling wine. The wily Louis Roederer Champagne house bypassed Napa and Sonoma to start its sparkling wine operation here and has done extremely well in a short time – as have Scharffenberger and Handley, two other successful sparkling wine producers in Anderson Valley.

San Francisco Bay Area

The San Francisco Bay area includes wine regions north, east, and south of the city: Marin County to the north; Alameda County and Livermore Valley to the east; and Santa Clara Valley and San Mateo County to the south.

The urban spread east and south of San Francisco, from the cities of Palo Alto to San Jose (Silicon Valley) and eastward, has taken its toll on vineyards in the Livermore and Santa Clara Valleys. These two growing regions, both cooled by breezes from the San Francisco Bay, are now relatively small.

In Livermore, directly east of San Francisco, Sauvignon Blanc and Sémillon have always done well. In Santa Clara Valley, south of San Francisco with the Santa Cruz Mountains on its western side, Chardonnay, Cabernet Sauvignon, and Merlot are the three big grape varieties (and wines).

Santa Cruz Mountains

Standing atop one of the isolated Santa Cruz Mountains, you can quickly forget that you're only an hour's drive south of San Francisco. The rugged, wild beauty of this area has attracted quite a few winemakers, including some of the best in the state. (Paul Draper of Ridge Vineyards and Randall Grahm of Bonny Doon are but two.) The climate is cool on the ocean side, where Pinot Noir thrives. On the San Francisco Bay side, Cabernet Sauvignon is the important red variety. Chardonnay is a leading variety on both sides.

What's New in Old Monterey

Monterey County has a little bit of everything – a beautiful coastline, the chic town of Carmel, some very cool (as in temperature, not chicness) vineyard districts and some very warm areas, mountain wineries and Salinas Valley wineries, a few gigantic wine firms and lots of small ones. Like most California wine regions, Monterey has been changing rapidly during the past two decades.

Chardonnay is the leading varietal wine in Monterey County – as it is in most of the state. But the cooler parts of Monterey are also principal sources of Riesling and Gewürztraminer. Cabernet Sauvignon, and Pinot Noir are the leading red varieties in the mountain areas.

Romancing in the Sierra Foothills

No wine region in America has a more romantic past than the Sierra Foothills. The Gold Rush of 1849 carved a place in history for the foothills of the Sierra Nevada Mountains. It also brought vineyards to the area to provide wine for the thirsty miners. One of the vines planted at that time was certainly Zinfandel – still the region's most famous wine. Many of the oldest grapevines in the United States, some over 100 years old – mainly Zinfandel – are here in the Sierra Foothills.

In fact, very little has changed in the Sierra Foothills over the years. This is clearly most rustic wine region on the West Coast – and perhaps in the country.

The Sierra Foothills is a sprawling wine region east of Sacramento, centered in Amador and El Dorado Counties, but spreading north and south of both. Two of its best-known viticultural areas are Shenandoah Valley and Fiddletown. Summers can be hot, but many vineyards are situated as high as 1,500 feet — such as around Placerville in El Dorado — and evenings are very cool. Soil throughout the region is mainly volcanic in origin.

Contrasts in San Luis Obispo

San Luis Obispo County is an area of vastly diverse viticultural areas. These include, for example, the warm, hilly Paso Robles region (north of the town of San Luis Obispo) where Zinfandel and Cabernet Sauvignon reign, and the cool, coastal Edna Valley and Arroyo Grande (south of the town), home of some very good Pinot Noirs and Chardonnays.

Paso Robles, with over 90 wineries, is in the heart of California's Central Coast, about equidistant from San Francisco and Los Angeles. Its wines are so different from those of the two coastal areas that we name the producers separately.

Santa Barbara, Californian Paradise

The most exciting viticultural areas in California – if not in the entire country – are in Santa Barbara County. The cool Santa Maria, Santa Ynez, and Los Alamos Valleys – which lie north of the city of Santa Barbara – run east to west, opening toward the Pacific Ocean and channeling in the ocean air. The cool climate is ideal for Pinot Noir and Chardonnay. In the Santa Maria Valley, one of the main sources of these varieties, the average temperature during the growing season is a mere 74°F. Farther south, in the Santa Ynez Valley, Riesling also does well.

Santa Barbara is generally recognized as one of the six great American wine regions for this variety. In Santa Barbara, Pinot Noir wines seem to burst with luscious strawberry fruit, laced with herbal tones. These wines tend to be precocious; they're delicious in their first four or five years – not the "keepers" that the sturdier, wilder-tasting Russian River Pinot Noirs seem to be. But why keep them when they taste so good?

Oregon, A Tale of Two Pinots

Because Oregon is north of California, most people assume that Oregon's wine regions are cool. And they're right. But the main reason for Oregon's cool climate is that no high mountains separate the vineyards from the Pacific Ocean. The ocean influence brings cool temperatures and rain. Grape growing and winemaking are really completely different in Oregon and California.

Oregon first gained respect in wine circles for its Pinot Noir, a grape that needs cool climates to perform at its best. Pinot Noir is still Oregon's flagship wine, and a vast majority of the state's wineries make this wine. Oregon's Pinot Noirs, with their characteristic black-fruit aromas and flavours, depth and complexity, have won accolades as among the very best Pinots in the United States.

Because Chardonnay is the companion grape to Pinot Noir in France's Burgundy region, and because Chardonnay wine is hugely popular in America, it's an important variety in Oregon. However, a second white grape variety has emerged to challenge Chardonnay's domination: Pinot Gris. A natural mutation of its ancestor, Pinot Noir, the Pinot Gris variety has grapes that are normally pale pink–yellowish in colour when ripe. Today, over 75 wineries in Oregon make Pinot Gris.

Two styles of Oregon Pinot Gris exist:

- A lighter, fruity style (for which the grapes are picked early) is always unoaked and can be consumed as soon as six to eight months after the autumn harvest.

- A medium-bodied, golden-coloured wine from grapes left longer on the vine sometimes has a little oak aging and can age for five or six years or longer.

In general, Oregon Pinot Gris is light- to medium-bodied, with aromas reminiscent of pears, apples, and sometimes of melon, and surprising depth for an inexpensive wine. It's an excellent food wine, even when it's slightly sweet; it works well especially with seafood and salmon, just the kind of food that it's paired with in Oregon. And the best news is the price. Most of Oregon's Pinot Gris wines are in the £7 to £11 range in retail stores.

Wine on the Desert: Washington State

Although Washington and Oregon are neighboring states, their wine regions have vastly different climates due to the location of the vineyards relative to the Cascade Mountains, which cut through both states from north to south.

On Washington's western, or coastal, side, the climate is maritime – cool, plenty of rain, and a lot of vegetation. (In Oregon, almost all the vineyards are located on the coastal side.) East of the mountains, Washington's climate is continental, with hot, very dry summers and cold winters. Most of Washington's vineyards are situated in this area, in the vast, sprawling Columbia and Yakima Valleys. Because it's so far north, Washington also has the advantage of long hours of sunlight, averaging an unusually high 17.4 hours of sunshine during the growing season.

Washington's winemakers have found that with irrigation, many grapes can flourish in the Washington desert. The Bordeaux varieties – Merlot, Cabernet Sauvignon, Cabernet Franc, Sauvignon Blanc, and Sémillon – are the name of the game. Syrah is coming up fast, and Chenin Blanc and the ever-present Chardonnay also are doing well.

Washington first became well-known for the quality of its Merlots. Lately, Washington's Syrah wines are gaining many of the accolades. In fact, Washington may be the single best region in the United States for this exciting wine. Cabernet Sauvignon and Cabernet Franc are also excellent varietal wines in Washington.

The Empire State

New York City may be the capital of the world in many ways, but its state's wines don't get the recognition they deserve, perhaps because of California's overwhelming presence in the U.S. market. New York ranks as the third largest wine producing state in the United States.

New York's most important region is the Finger Lakes, where four large lakes temper the otherwise cool climate. This AVA produces about two-thirds of New York's wines. The other two important regions are the Hudson Valley, along the Hudson River north of New York City, and Long Island.

Like Washington state, Long Island seems particularly suited to Merlot, but Chardonnay, Riesling, Cabernet Sauvignon, Cabernet Franc, and Sauvignon Blanc are also grown, plus some Gewürztraminer, Pinot Noir, and numerous other varieties.

Oh, Canada

Canada's wines are known mainly to Canadians, who consume the bulk of their country's production. The 1990s brought incredible growth to the Canadian wine industry. Wine is made in four of Canada's provinces, but Ontario has bragging rights as the largest producer, with over 100 wineries. British Columbia ranks second. Quebec and Nova Scotia also produce wine.

To identify and promote wines made entirely from local grapes (some Canadian wineries import wines from other countries to blend with local production), the provinces of Ontario and British Columbia have established an appellation system called VQA, Vintners' Quality Alliance. This system regulates the use of provincial names on wine labels, establishes which grape varieties can be used (vinifera varieties and certain hybrids), regulates the use of the terms icewine, late harvest, and botrytised and requires wines to pass a taste and laboratory test.

Ontario

Ontario's vineyards are cool-climate wine zones, despite the fact that they lie on the same parallel as Chianti Classico and Rioja, warmer European wine regions. Sixty per cent of the production is white wine, from Chardonnay, Riesling, Gewürztraminer, Pinot Blanc, Auxerrois, and the hybrids Seyval Blanc and Vidal. Red wines come from Pinot Noir, Gamay, Cabernet Sauvignon, Cabernet Franc, Merlot, and the hybrids Maréchal Foch and Baco Noir.

TIP

Because winter temperatures regularly drop well below freezing, *icewine,* made from grapes naturally frozen on the vine, is a specialty of Ontario. It is gradually earning the Canadian wine industry international attention, particularly for the wines of Inniskillin Winery. VQA regulations are particularly strict regarding icewine production, as it has developed into the leader of the Canadian wine exports.

British Columbia

The rapidly growing wine industry of British Columbia now boasts more than 70 wineries. Production is mainly white wine – from Chardonnay, Gewürztraminer, Pinot Gris, Pinot Blanc, and Riesling – but red wine production is increasing, mainly from Pinot Noir and Merlot.

The Okanagan Valley in southeast British Columbia, where the climate is influenced by Lake Okanagan, is the center of wine production.

Does Wine Really Breathe?

Most wine is alive in the sense that it changes chemically as it slowly grows older. Wine absorbs oxygen and, like your own cells, it oxidizes. When the grapes turn into wine in the first place, they give off carbon dioxide. So you could say that wine breathes, in a sense.

If you really want to aerate your wine, do one or both of the following:

- Pour the wine into a decanter.
- Pour the wine into large glasses at least ten minutes before you plan to drink it.

Many red wines but only a few white wines – and some dessert wines – can benefit from aeration. You can drink most white wines upon pouring, unless they're too cold.

WARNING!

Practically speaking, it doesn't matter what your decanter looks like or how much it costs. In fact, the very inexpensive, wide-mouthed carafes are fine.

TIP

The term *breathing* refers to the process of aerating the wine, exposing it to air. Sometimes the aroma and flavour of a very young wine will improve with aeration. But just pulling the cork out of the bottle and letting the bottle sit there is a truly ineffective way to aerate the wine. The little space at the neck of the bottle is way too small to allow your wine to breathe very much.

Young, tannic red wines

The younger and more tannic the wine is, the longer it needs to breathe. As a general rule, most tannic, young red wines soften up with one hour of aeration.

TIP

Young, tannic red wines – such as Cabernet Sauvignons, Bordeaux, many wines from the northern Rhône Valley, and many Italian wines – actually taste better with aeration because their tannins soften and the wine becomes less harsh.

Older red wines with sediment

Many red wines develop *sediment* (tannin and other matter in the wine that solidifies over time) usually after about eight years of age. You'll want to remove the sediment because it can taste a bit bitter. Also, the dark particles floating in your wine, usually at the bottom of your glass, don't look very appetizing.

To remove sediment, keep the bottle of wine upright for a day or two before you plan to open it so that the sediment settles at the bottom of the bottle. Then decant the wine carefully: Pour the wine out of the bottle slowly into a decanter while watching the wine inside the bottle as it approaches the neck. You watch the wine so that you can stop pouring when you see cloudy wine from the bottom of the bottle making its way to the neck. If you stop pouring at the right moment, all the cloudy wine remains behind in the bottle.

If the wine needs aeration after decanting (that is, it still tastes a bit harsh), let it breathe in the open decanter. If the wine has a dark colour, chances are that it is still quite youthful and needs to breathe more. Conversely, if the wine has a brick red or pale garnet colour, it probably has matured and may not need much aeration.

A few white wines

Some very good, dry white wines – such as full-bodied white Burgundies and white Bordeaux wines, as well as the best Alsace whites – also get better with aeration. For example, if you open a young Corton-Charlemagne (a great white Burgundy), and it doesn't seem to be showing much aroma or flavour, chances are that it needs aeration. Decant it and taste it again in half an hour. In most cases, the wine dramatically improves.

Vintage Ports

One of the most famous fortified wines is Vintage Port (properly called "Porto"). For now, just say that, yes, Vintage Port needs breathing lessons, and needs them very much indeed! Young Vintage Ports are so brutally tannic that they demand many hours of aeration (eight would not be too many). Even older Ports improve with four hours or more of aeration. Older Vintage Ports require decanting for another reason: They're chock-full of sediment. (Often, large flakes of sediment fill the bottom 10 per cent of the bottle.) Keep Vintage Ports standing for several days before you open them.

Australian and New Zealand Wines Arise

IN THIS ARTICLE

- *Comparing Old World to New World*
- *Sizzling with Australian wines*
- *Exploring new wines from New Zealand*

What do the wines of North and South America, South Africa, Australia, and New Zealand have in common? For one thing, none of them are produced in Europe. In fact, you could say that they are the wines of "Not Europe."

The name most often used in wine circles for Not Europe is the *New World*. Undoubtedly this phrase, with its ring of colonialism, was coined by a European. Europe, home of all the classic wine regions of the world, producer of more than 60 per cent of the world's wine, is the Old World. Everything else is nouveau riche.

This chapter explores the wines of Australia and New Zealand, while the next chapters talks about the wines of Chile, Argentina, and South Africa.

Old World Versus New World

The expression *New World,* when applied to wines, lumps together wine regions as remote as Napa Valley, the Finger Lakes, Coonawarra, and Chile's Maipo Valley.

The New World is a winemaking entity whose legislative reality, spirit, and winemaking style are unique from those of the Old World – as generalizations go.

In wine terms, the New World isn't just geography but also an attitude toward wine. Some winemakers in Europe approach wine the liberated New World way, and some winemakers in California are dedicated Old World traditionalists. Keep that in mind as you look over the following comparison between the Old and the New. And remember, these are generalizations– and generalizations are never always true.

New World	Old World
Innovation	*Tradition*
Wines named after grape varieties	Wines named after region of production
Expression of the fruit is the primary winemaking goal	Expression of the terroir (the particular place where the grapes grow, with its unique growing conditions) is the winemaking goal
Technology is revered	Traditional methods are favored
Wines are flavorful and fruity	Wines have subtle, less fruity flavors
Grape-growing regions are broad and flexible	Grape-growing regions are relatively small and fixed
Winemaking resembles science	Winemaking resembles art
Winemaking processes are controlled	Intervention in winemaking is avoided as much as possible
The winemaker gets credit for the wine	The vineyard gets the credit

Australian Wine Power

Make no mistake about it: Australia is one of the world powers of wine. The wine industry of Australia is perhaps the most technologically advanced, forward-thinking on earth, and the success of Australian wines around the world is the envy of wine producers in many other countries.

Australia has no native vines. Vinifera grapevines first came to the country in 1788, from South Africa. At first, most Australian wines were rich and sweet, many of them fortified, but today Australia is famous for its fresh, fruity red and white table wines that manage to be extremely consistent in quality. Australia now ranks sixth in the world in wine production — making slightly more than half as much wine as the United States — and fourth in exports.

Approximately the same size as the continental United States, Australia has about 2,000 wineries. Many of these wineries are small, family-owned companies, but four mega-companies — Foster's Wine Group, Constellation Wines, Pernod Ricard, and McGuigan Simeon Wines — together with one family-owned winery, Casella Wines, are responsible for about two-thirds of Australia's wine production.

Winemaking, grapes, and terroir

Australia's wine regions are mainly in the southern, cooler part of the country, with many of them clustered in the state of Victoria, the southern part of South Australia, the southern part of Western Australia, and the cooler parts of New South Wales.

The success of Australia's wines stems from a generally warm, dry climate, which provides winemakers with excellent raw material for their work. The country's research programs in grape-growing and winemaking also contribute greatly by enabling winemakers to stay on the cutting edge of their craft.

Australia's number-one grape for fine wine is Syrah, locally called *Shiraz,* followed by Cabernet Sauvignon, Chardonnay, Merlot, Semillon (pronounced *SEM eh lon* in Australia, as opposed to the French *sem ee yon* elsewhere in the world), Pinot Noir, Riesling, and Sauvignon Blanc. The wines are generally labeled with the name of their grape variety, which must constitute at least 85 per cent of the wine.

TIP

Shiraz wines are particularly interesting because they come in numerous styles, from inexpensive, juicy wines brimming with ripe plum and blackberry fruit to serious wines that express specific regional characteristics, such as spice and pepper from cool-climate areas (such as Yarra Valley and the Adelaide Hills) or sweet-fruit ripeness from warmer areas (such as McLaren Vale, Barossa, and Clare).

The wines of Australia have two distinct faces:

- Most Australian wines in export markets are inexpensive varietal wines that sell for £6 a bottle or less. These wines are generally labeled simply as coming from *South Eastern Australia,* meaning that the grapes could have come from any of three states, a huge territory. Often sporting whimsical labels, they are user-friendly wines that preserve the intense flavours of their grapes and are soft and pleasant to drink young.
- Higher-priced wines carry more focused regional designations, such as single states (South Australia or Victoria, for example) or even tighter region-specific designations (such as Coonawarra or Yarra Valley). Although these wines are also enjoyable when released, they are more serious wines that can also age. Australia now has 60 wine regions and more than 100 Geographic Indications (GIs).

Odd couples

Although winemakers all over the world make blended wines – wines from more than one grape variety – generally the grape combinations follow the classic French models: Cabernet Sauvignon with Merlot and Cabernet Franc, for example, or Sémillon with Sauvignon Blanc. Australia has invented two completely original formulas:

- Shiraz with Cabernet Sauvignon
- Semillon with Chardonnay

The grape in the majority is listed first on the wine label for wines sold in the United States, and the percentages of each grape are indicated.

Australia's wine regions

Australia's most important state for wine production is South Australia, whose capital is Adelaide (see figure). South Australia makes about 50 per cent of Australia's wine. While many vineyards in South Australia produce inexpensive wines for the thirsty home market, vineyards closer to Adelaide make wines that are considered among the country's finest. Among these fine wine regions are

- **Barossa Valley:** North of Adelaide, this is one of Australia's oldest areas for fine wine; it's a relatively warm area famous especially for its robust Shiraz, Cabernet Sauvignon, and Grenache, as well as rich Semillon and Riesling (grown in the cooler hills). Most of Australia's largest wineries, including Penfolds, are based here.
- **Clare Valley:** North of the Barossa Valley, this climatically diverse area makes the country's best Rieslings in a dry, weighty yet crisp style, as well as fine Shiraz and Cabernet Sauvignon.
- **McLaren Vale:** South of Adelaide, with a mild climate influenced by the sea, this region is particularly admired for its Shiraz, Cabernet, Sauvignon Blanc, and Chardonnay.
- **Adelaide Hills:** Situated partially within the Adelaide city limits, this fairly cool region sits between the Barossa and McLaren Vale areas and is the home to rather good Sauvignon Blanc, Chardonnay, Pinot Noir, and Shiraz.

The wine regions of Australia.

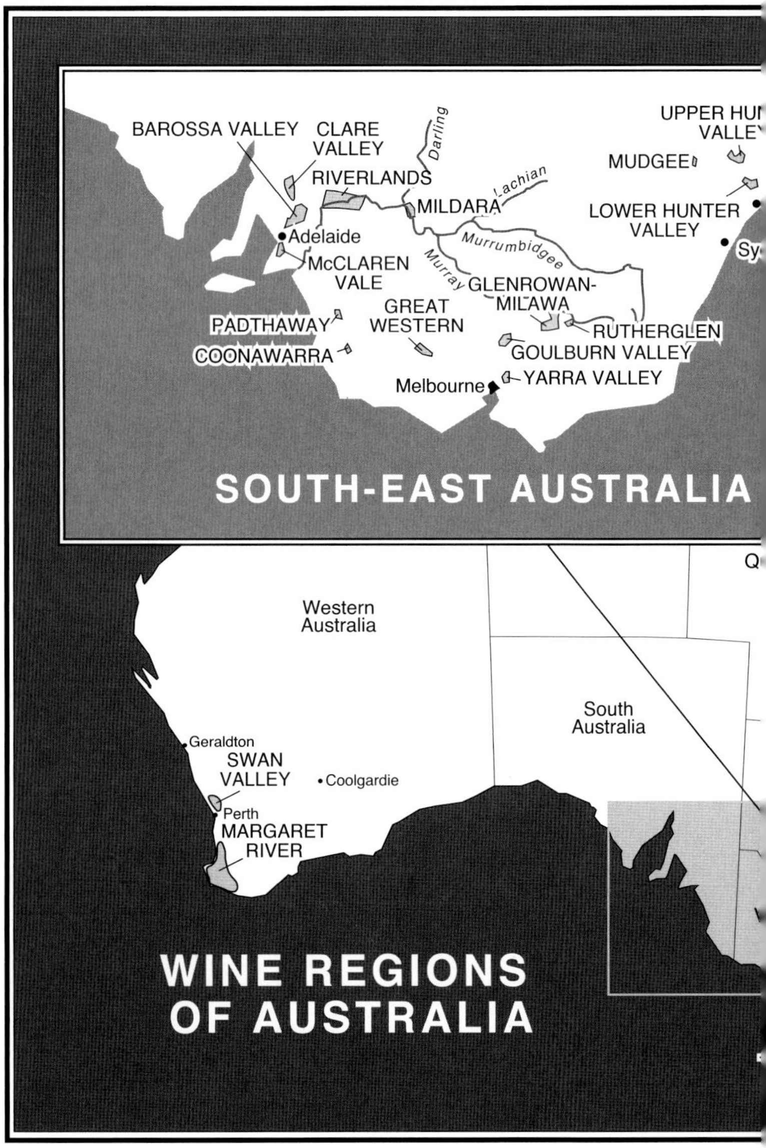

- **Limestone Coast:** This unique zone along the southern coast of South Australia is an important area for fine wine, both red and white, thanks to the prevalence of limestone in the soil. Two of the six regions within the Limestone Coast zone are famous in their own right — the cool **Coonawarra** for some of Australia's best Cabernet Sauvignon wines, and **Padthaway** for its white wines, particularly Chardonnay, Sauvignon Blanc, and Riesling.

Adjoining South Australia to the east is Victoria, a smaller state that makes 15 per cent of Australia's wines. While South Australia is home to most of Australia's largest wineries, Victoria has more wineries (over 500), most of them small. Victoria's fine wine production ranges from rich, fortified dessert wines to delicate Pinot Noirs. Principal regions include, from north to south

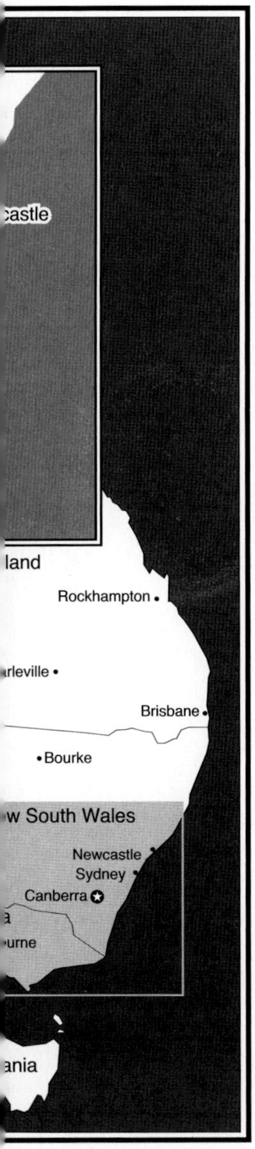

- **Murray River:** This area stretching into New South Wales includes the Mildura region, where Lindemans, one of Australia's largest wineries is situated. This region is particularly important for growing grapes for Australia's good-value wines.

- **Rutherglen:** In the northeast, this long-established, warm climate zone is an outpost of traditional winemaking and home of an exotic Australian specialty, fortified dessert Muscats and Tokays.

- **Goulburn Valley:** In the center of the state, Goulburn Valley is known especially for its full-bodied reds, especially Shiraz.

- **Heathcote:** East of Goulburn and due north of Melbourne (the capital), this area boasts unusual soils that make distinctive, rich-yet-elegant Shirazes and also Cabernet.

- **Yarra Valley:** In southern Victoria, and close to Melbourne, Yarra Valley boasts a wide diversity of climates due to altitude differences of its vineyards. The Yarra is noted for its Cabernet, Pinot Noir, Shiraz, Chardonnay, and Sauvignon Blanc.

- **Mornington Peninsula and Geelong:** South of Melbourne and separated from each other by Port Phillip Bay, these two cool, maritime regions specialize in fine Pinot Noir and Chardonnay.

New South Wales, with its capital, Sydney, is Australia's most populous state, and the first to grow vines; today it makes 31 per cent of Australia's wine. High-volume production of everyday wines comes from an interior area called the Riverina. Fine wine, for now, comes from three other areas:

- **Hunter Valley:** An historic grape-growing area that begins 80 miles north of Sydney. The Lower Hunter, with a warm, damp climate and heavy soils, produces long-lived Semillon as its best wine. The Upper Hunter is a drier area farther from the coast.

- **Mudgee:** An interior area near the mountains. Mudgee specializes in reds such as Merlot and Cabernet Sauvignon but also makes Chardonnay.

- **Orange:** A cool, high-altitude area making distinctive white wines and also very good reds.

Western Australia, the country's largest state, with its most isolated wine area – in the southwest corner – makes little wine compared to the preceding three states, but quality is high. The warm, dry Swan Valley is the state's historic center of wine production, but two cooler climate regions have become more important:

- **Margaret River:** This is a relatively temperate region near the Indian Ocean. Among the wines that various wineries here excel in are Sauvignon Blanc-Semillon blends (especially Cape Mentelle), Chardonnay (especially Leeuwin Estate), and Cabernet Sauvignon (from Mosswood, Voyager, Cape Mentelle, and Howard Park).

- **Great Southern:** Cooler than Margaret River, Great Southern's specialty is crisp, age-worthy Riesling. This huge, diverse region produces intense, aromatic Cabernet Sauvignon as well as fine Shiraz and Chardonnay; on the southern coast, Pinot Noir is successful.

Tasmania has some cool microclimates where producers such as Pipers Brook are proving what potential exists for delicate Pinot Noirs, Chardonnays, and sparkling wines.

The Rise of New Zealand

The history of fine winemaking in New Zealand is relatively short, having been hampered by conservative attitudes towards alcohol. In the 1980s, New Zealand finally began capitalizing on its maritime climate, ideal for producing high-quality wines, and started planting grapes in earnest. Today, it makes less than one-tenth of the wine of its nearest neighbour, Australia, but its production is increasing every year. And, unlike Australia, New Zealand has managed to maintain an elite image for its wines, as opposed to a good-value-for-everyday image.

Situated farther south than Australia, New Zealand is, in general, cooler. Of New Zealand's two large islands, the North Island is the warmer. Red grapes grow around the capital city, Auckland, in the north and around Hawkes Bay (especially known for its Cabernet Sauvignon) farther south on the North Island; Müller-Thurgau, Chardonnay, and Sauvignon Blanc are that island's main white varieties. Martinborough, a cooler district at the southern end of North Island, makes very good Pinot Noir.

On the South Island, Marlborough – the country's largest and commercially most important wine region – is New Zealand's top production zone for Chardonnay and, especially, Sauvignon Blanc.

The first New Zealand Sauvignon Blancs to be exported were generally unoaked wines with pronounced flavour, rich texture, and high acidity. They were so distinctive – pungent, herbaceous, with intense flavours suggestive of asparagus, lime, or cut grass – that New Zealand became recognized almost overnight in the late 1980s for a new prototype of Sauvignon Blanc. This style of New Zealand Sauvignon Blanc is still very popular worldwide. These are the least expensive Kiwi Sauvignon Blancs, retailing for £7 to £11, with many priced around £9.

Another style of New Zealand Sauvignon Blanc has evolved in the last decade. Riper, less assertive, and softer in texture, this style is often achieved through the use of oak barrels and/or blending with Semillon, and it has fruitier flavours, usually passion fruit or ripe grapefruit. New Zealand wine producers correctly foresaw that wine drinkers may need an alternative to the herbaceous style. The riper, fruitier, less herbaceous New Zealand Sauvignon Blancs are frequently labeled as "Reserve" wines or as single-vineyard wines. They generally retail for £11 to £18.

Pinot Noir is increasingly significant in New Zealand. In addition to its stronghold in Martinborough, on the North Island, Pinot Noir is now being made in Marlborough and throughout the South Island, and this grape has now surpassed Cabernet Sauvignon as New Zealand's most planted

red variety. New Zealand Pinot Noirs vary in taste from region to region; the wines of Martinborough, for example, are a bit more savory and minerally than those of Marlborough, which tend to be soft and fruity. In time, as the producers of each region refine their styles, the regional differences should become more evident.

Four large producers dominate New Zealand's wine production: Montana, Corbans, Villa Maria, and Nobilo. But in the past 20 years, numerous small, boutique wineries have sprung up, especially on the South Island, and are making excellent wine.

TIP

In the central part of the South Island, Central Otago, home of the world's most southerly grapevines, has emerged as one of New Zealand's top regions for Pinot Noir. Vines are planted on hillsides for more sunshine and less risk of frost. The low-yielding vines here produce highly-concentrated Pinot Noir wines. Mt. Difficulty and Felton Road, both from Central Otago, are currently two of New Zealand's best Pinot Noir producers.

Current trends in Kiwi Land

New Zealand Sauvignon Blancs are still hot and Pinot Noir seems to be the next Big Thing. But New Zealand is more than just a two-grape country. In the white wine category, the improved Chardonnays, Rieslings, and Pinot Gris wines are impressive. The biggest surprise could be New Zealand's really fine Cabernet Sauvignons, Merlots, and Bordeaux-blends, not only from established warmer-climate North Island regions such as Hawke's Bay and its Gimblett Road zone, but also from Waiheke Island, a few miles east of the city of Auckland, where the climate is mild enough to grow Cabernet Franc and Petit Verdot.

New Zealand's final surprise is that it's making excellent sparkling wines by using the classic method. Most of the better New Zealand sparkling wines also use the two main grape varieties of Champagne, Pinot Noir and Chardonnay. Highfield Estate is one of New Zealand's many fine sparkling wine producers.

Heating It Up with Chile, Argentina, and South Africa

IN THIS ARTICLE

- *Discovering Chile's wines*
- *Boasting about Argentina's wines*
- *Exploring South Africa's wines*

In Europe, they've been making wines for so long that grape-growing and winemaking practices are now codified into detailed regulations. Which hillsides to plant, which grapes should grow where, how dry or sweet a particular wine should be – these decisions were all made long ago, by the grandparents and great-great-grandparents of today's winemakers. But in New World wine countries, the grape-growing and winemaking game is wide open; every winery owner gets to decide for himself where to grow his grapes, what variety to plant, and what style of wine to make.

This article explores the wines of Chile, Argentina, and South Africa.

Chile Discovers Itself

Chile's wine industry wears the mantle *New World* somewhat uncomfortably. The Spanish first established vineyards in Chile in the mid-sixteenth century, and the country has maintained a thriving wine industry for its home market for several centuries. Nothing new about that. What *is* new about Chile, however, is the growth of her wine industry since the mid-1980s, her rapid development of a strong export market, and her shift toward French grape varieties such as Cabernet Sauvignon, Merlot, and Chardonnay – with an almost-forgotten red Bordeaux variety called Carmenère definitely in the running on the outside.

With the Pacific Ocean to the west and the Andes Mountains to the east, Chile is an isolated country. This isolation has its advantages in terms of grape growing: the vine-destroying louse *Phylloxera* hasn't yet taken hold in Chile – as it's done in just about every other winemaking country – and vinifera vines can therefore grow on their own roots. Chile's other viticultural blessings include a range of mountains along the coast, which blocks the ocean dampness from most vineyards, and the ocean's general tempering influence on a relatively hot climate.

Chile's wine regions

As in every other country, grape growers and wine producers in Chile originally planted vineyards in the most obvious locations, where grapes would grow prolifically. Trial and error have gradually enabled them to discover the less obvious locations — many of them cooler and less accessible areas — that offer the opportunity to make truly distinctive wines.

Initially considered the ideal place to plant grapes, Maipo Valley is part of Chile's vast Central Valley, which lies between the coastal range and the Andes. Convenience played a large role: The Maipo Valley surrounds Santiago, Chile's capital and its largest and most important city. Most of Chile's vineyards are still in the Central Valley, but today, vineyards also exist in regions that no one had heard of just ten years ago.

From north to south, here's a summary of Chile's wine regions today, both old and new:

- **Limari Valley:** A small region northwest of Santiago, near the Pacific Ocean. Although the climate is hot and dry — it's nearer to the equator than any of Chile's other important regions and close to the Atacama Desert — its unique microclimate, caused by its proximity to the Pacific, features cooling morning fog and ocean breezes that blow through the Valley during the day. Chile's three largest wineries, Concha y Toro, San Pedro, and Santa Rita, all have bought land in Limari. Promising wines so far are Sauvignon Blanc, Chardonnay, and Syrah. This is one of the country's hot emerging regions.

- **Aconcagua Valley:** North of Santiago, Aconcagua Valley is named after South America's highest mountain, the magnificent Mount Aconcagua, and is one of the warmest areas for fine grapes. But Aconcagua also includes many cooler high-altitude sections. Cabernet Sauvignon grows especially well here, and more recently, Syrah. Viña Errázuriz is Aconcagua Valley's most important winery.

- **Casablanca Valley:** Once considered part of the Aconcagua Valley, the cooler Casablanca valley, near the Pacific Ocean, now has its own identity. The first-established of the newer Chilean wine regions, it's still one of the best. Some of Chile's finest Chardonnays and Sauvignon Blancs grow in one part of Casablanca, while good Merlots and Pinot Noirs come from a more mountainous part. Veramonte is Casablanca's best-known winery, but many other wineries own vineyards in this region.

- **San Antonio Valley:** Along with nearby Leyda Valley, tiny San Antonio Valley, south of Casablanca Valley and next to the ocean, is arguably Chile's most exciting new region. Pinot Noir and Syrah are growing especially well on its cool, steep slopes. Now making one of the world's best Pinot Noirs outside of Burgundy and a fine Syrah, Viña Matetic is the winery to watch in San Antonio Valley.

- **Maipo Valley:** Chile's most established wine region, just south of Santiago, Maipo Valley is home to most of the country's wineries. Concha y Toro, Santa Rita, and Almaviva are a few of Maipo's premium producers. Cabernet Sauvignon is king in this region, and Merlot also does very well.

- **Cachapoal Valley:** The large Rapel Valley, south of Maipo Valley, has two main wine regions, Cachapoal Valley and Colchagua Valley. Cachapoal Valley, nearer the Andes, is a red wine region, and is strong in Merlot and Cabernet Sauvignon. Morandé and Altair are two rising star wineries here.

- **Colchagua Valley:** Ocean breezes have transformed the formerly quiet Colchagua Valley into one of Chile's most important new red wine regions. Carmenère, Cabernet Sauvignon, Merlot, and Syrah grow especially well here. Colchagua's two leading wineries are Casa Lapostolle and Montes.

- **Curicó Valley:** One of Chile's oldest and largest wine regions, the Curicó Valley is directly south of Rapel Valley. Because of its diverse microclimates, both red and white varieties grow well here. The huge San Pedro Winery and Viña Miguel Torres are located in Curicó.

- **Maule:** Maule Valley is Chile's largest wine region in area, and also the southernmost of its important wine regions. Because it's so huge, it has many diverse microclimates, and both red and white varieties grow well, especially Sauvignon Blanc, Cabernet Sauvignon, and Merlot. Viña Calina is Maule Valley's best-known winery.

The face and taste of the wines

Stylistically, Chile's wines generally lack the exuberant fruitiness of Californian and Australian wines. And yet they're not quite as subtle and understated as European wines. Although red wines have always been Chile's strength, today the white wines, especially those from cooler regions, are very good. Chile's Sauvignon Blancs are generally unoaked, while most of the Chardonnays are oaked.

TIP

Like most New World wines, Chile's wines are generally named after their grape varieties; they carry a regional (or sometimes a district) indication, too. The reasonable prices of the basic wines – mainly from £3 to £6 – make these wines excellent value. The most important wineries for the export market include, in alphabetical order, Calina, Caliterra, Carmen, Casa Lapostolle, Concha y Toro, Cousiño Macul, Errazuriz, Haras de Pirque, Los Vascos, Montes, Mont Gras, Santa Carolina, Santa Rita, and Undurraga. Viña Matetic, a rising star from the San Antonio Valley, is just beginning to appear on the export markets.

Chile's new challenge is to produce good-quality high-end wines along with its inexpensive varietals. Many of the top producers now make a super-premium red wine in the £28 to £56 price range. These elite Chilean reds are often blends rather than varietal wines, and many are styled along international lines — made from very ripe grapes that give rich, fruity flavours and high (14 per cent or higher) alcohol levels, and aged in small French oak barrels. What many (but not all) of them lack, however, is a sense of place: They don't taste particularly Chilean. With time, Chile will undoubtedly reach its goal and begin producing fine wines that merit their high prices.

Keep an eye out for some of Chile's top super-premium red wines:

- Concha y Toro's Don Melchor Cabernet Sauvignon (about £28)
- Errázuriz's Don Maximiano Founder's Reserve (mainly Cabernet Sauvignon, about £31)
- Albis, a wine from a joint venture between the Chilean Haras de Pirque winery and the Italian Antinori company (Cabernet Sauvignon and Carmenère, about £32)
- Montes Alpha M (a "Bordeaux blend," about £47)
- Almaviva (a sleek and subtle red, mainly Cabernet Sauvignon with Carménère and Cabernet Franc, about £56)
- Casa Lapostolle's Clos Apalta (a blend of Carménère, Merlot, and Cabernet Sauvignon, about £50)
- Seña, from an estate in Aconcagua that was originally a partnership between the Robert Mondavi and Eduardo Chadwick (of Viña Errázuriz) families and is now owned by the Chadwicks (Cabernet Sauvignon, Merlot, and Carmenère, about £40)

Argentina, a Major League Player

Argentina produces about four times as much wine as Chile does — almost as much as the entire United States. It boasts the largest wine production in South America and the fifth-largest wine production in the world. In recent years, winemaking has shifted away from large-volume wines suited to the domestic market and toward higher-quality wines that suit wine drinkers outside Argentina. Not only is Argentina now a major player in the world wine market, but it's one of the world's most exciting countries for wine production.

Regions and grapes

Argentina's wine regions are situated mainly in the western part of the country, where the Andes Mountains divide Argentina from Chile. High altitude tempers the climate, but the vineyards are still very warm by day, cool by night, and desert dry. Rivers flow through the area from the Andes and provide water for irrigation.

The vast majority of Argentina's vineyards are in the state of Mendoza, Argentina's largest wine region, which lies at roughly the same latitude as Santiago, Chile. Within the Mendoza region are various wine districts (the names of which sometimes appear on wine labels) such as Maipú, San Martín, Tupungato, and Luján de Cuyo. Most of Argentina's oldest wineries and their vineyards are clustered close to Mendoza city, but the Uco Valley, south of the city, has attracted many newcomers who are building impressive wineries.

San Juan, just north of Mendoza and considerably hotter, is Argentina's second-largest wine region. La Rioja, Argentina's oldest wine-producing region, is east of San Juan.

TIP

San Juan is particularly famous for Torrontés, a variety that's probably indigenous to Galicia, Spain. It produces an inexpensive (£3 to £6), light-bodied, high-acid, aromatic white wine that's one of Argentina's signature white wines. It's especially fine with appetizers, seafood, and fish.

Argentina's red wines are generally higher in quality than its whites. The little-known Malbec grape variety — now seldom used in Bordeaux, where it originated — has emerged as Argentina's flagship variety. Malbec has adapted extremely well to the Mendoza region, and winemakers are learning how it varies in Mendoza's subzones. Arguments continue as to which variety makes Argentina's greatest red wines, Cabernet Sauvignon or Malbec. But the fact remains that good Cabernet wines come from almost every wine-producing country; only Argentina and Cahors, a small region in Southwest France, have had success with Malbec. The same logic suggests that Bonarda and Barbera, two northern Italian varieties that are widely planted in Argentina (especially Bonarda), have a good future there.

Names to know

Thanks in part to its high altitudes and sunny days, Argentina's natural resources for grape growing are among the strongest in the world. Increasingly, foreign investment continues to bring the capital and the winemaking know-how to make the most of these natural resources. Bodega Norton, for example, a winery that was purchased by an Austrian crystal producer in 1989, now makes some of the country's best wines. Moët & Chandon, another immigrant, is already Argentina's largest sparkling wine producer; it also makes the Terrazas varietal table wines. A Dutchman owns the state-of-the-art Bodegas Salentein winery and its sister winery, Finca El Portillo. Kendall Jackson has a presence, with its Viña Calina, as do several Bordeaux producers, such as Bordeaux's Lurton family, which owns Bodega J & F Lurton.

French enologist Michel Rolland has worked wonders at Trapiche; try Trapiche's great-value Oak Cask Cabernet Sauvignon or Oak Cask Malbec, both about £6. In fact Rolland, who knows the world's vineyards as well as anyone, has personally invested in an Argentine winery, Clos de los Siete; *that* says something about Argentina's potential!

TIP

The homegrown Catena Zapata has emerged as one of Argentina's top wine producers. At £6 a bottle, its Alamos Malbec is one of the greatest wine values around. Catena Cabernet Sauvignon or Malbec (both about £13), and the super-premium Malbec Alta or Cabernet Sauvignon Alta, both about £31, are higher-end wines, among the finest being made in South America today.

TIP

Other recommended Argentine producers include Bodega Norton, Bodega J. & F. Lurton, Bodegas Salentein, Bodega Weinert, Trapiche, Etchart, Finca Sophenia, Achaval Ferrer, Pascual Toso, Michel Torino, Las Terrazas, Navarro Correas, Santa Julia, El Portillo, Dona Paula, and Valentín Bianchi. Some of Argentina's basic wines are priced the same as Chile's, in the £3 to £6 range, but a few wineries make pricier wines starting in the £11 to £12 range.

South African Wine Safari

Despite being separated from Europe by nearly 3000 miles, South Africa's wines are rather reminiscent of European wines. The taste of a South African Cabernet Sauvignon, for example, may remind you of a French wine – but not quite. On the other hand, it doesn't really resemble a New World red from California or Australia, either. South African wines manage to combine the subtlety and finesse of French wines along with a touch of the voluptuous ripeness of California wines. In short, they are somewhat between both worlds.

Most of South Africa's table wines come from an area known as the Coastal Region, around the Cape of Good Hope. Traditionally, large firms dominated South Africa's wine industry, and they continue to do so. KWV, formerly a wine growers' cooperative, is one of the country's largest wineries. South Africa's largest winery, the gigantic Distell firm, owns two groups of wineries that had been among the country's largest wine companies — Stellenbosch Farmers' Winery Group and the Bergkelder Group.

South Africa's principal wine regions

South Africa has some vineyard areas with cool microclimates, especially around the southern coast (near the Cape of Good Hope) and in higher altitudes, but the climate in most of its wine regions is warm and dry.

South Africa's Wine of Origin legislation in 1973 created various wine regions, districts and wards. Almost all the country's vineyards are near its southwestern coast, in Cape Province, within 90 miles of Cape Town, the country's most fascinating and picturesque city.

The five major districts – mainly in the Coastal Region area – are

- **Constantia:** The oldest wine-producing area in the country (located south of Cape Town)
- **Stellenbosch:** East of Cape Town; the most important wine district in quantity and quality
- **Paarl:** North of Stellenbosch; home of the KWV and the famous, beautiful Nederburg Estate; the second-most important wine district
- **Franschhoek Valley:** A subdistrict of Paarl; many innovative winemakers here
- **Robertson:** East of Franschhoek, the only major district not in the Coastal Region; a hot, dry area, known mainly for its Chardonnays

The small, cool Hermanus/Walker Bay area, bordering the Indian Ocean, is also showing promise with Pinot Noir and Chardonnay, led by the innovative Hamilton Russell Winery. A newly added (11th) wine district, Elgin, is on the coast between Stellenbosch and Walker Bay. A cool area, Elgin shows promise for its intensely flavoured Sauvignon Blancs and for Pinot Noirs. The latest area to show promise is Darling Hills, north of Cape Town, led by an up-and-coming winery, Groote Post.

Steen, Pinotage, and company

The most-planted grape variety in South Africa is Chenin Blanc, often locally called *Steen*. This versatile grape primarily makes medium-dry to semi-sweet wines, but also dry wines, sparkling wines, late harvest botrytis wines, and rosés.

Cabernet Sauvignon, Merlot, Shiraz, and Pinot Noir have become increasingly important red varieties, while Sauvignon Blanc and Chardonnay are popular white varieties. Cabernet Sauvignon and Sauvignon Blanc do particularly well in South Africa's climate. (Producers here make a very assertive version of Sauvignon Blanc.)

And then you have Pinotage. Uniquely South African, Pinotage is a grape born as a crossing between Pinot Noir and Cinsaut (the same as Cinsault, the Rhône variety) back in 1925. However, Pinotage didn't appear as a wine until 1959. Pinotage wine combines the cherry fruit of Pinot Noir with the earthiness of a Rhône wine. It can be a truly delicious, light- to medium-bodied red wine that makes for easy drinking, or a more powerful red. Although many good Pinotage wines sell for £7 to £10, the best Pinotages cost more. Kanonkop Estate, a specialist in this variety, makes a £18 Pinotage. Simonsig Estate makes a fine Pinotage for £8.

While Pinotage is a pleasant wine, certainly worth trying, South Africa's future is with Cabernet Sauvignon, Merlot, and Shiraz (and blends of these grapes) for its red wines and Sauvignon Blanc and Chardonnay for its whites.

Champagnes and Other Sparkling Wines

IN THIS ARTICLE

- *Discovering when extra dry means "not all that dry"*
- *Realizing that all champagne is not Champagne*
- *Finding sparkling wines from £5 to £125+*

In the universe of wine, sparkling wines are a solar system unto themselves. They're produced in just about every country that makes wine, and they come in a wide range of tastes, quality levels, and prices. Champagne, the sparkling wine from the Champagne region of France, is the brightest star in the sky, but by no means the only one.

In many wine regions, sparkling wines are just a sideline to complement the region's table wine production, but in some places, sparkling wines are serious business. At the top of that list is France's Champagne region (where sparkling wine was – if not invented – made famous). Italy's Asti wine zone is another important region, as are France's Loire Valley, northeastern Spain, and parts of California. Australia, New Zealand, and South Africa are also now making some interesting sparklers.

Sparkling Wine Styles

All sparkling wines have bubbles, and nearly all of them are either white or pink (which is far less common than white). That's about as far as broad generalizations take you in describing sparkling wines.

Some sparkling wines are downright sweet, some are bone dry, and many fall somewhere in the middle, from medium-dry to medium-sweet. Some have toasty, nutty flavours, while some are fruity. Among the fruity sparkling wines, some are just nondescriptly grapey, while others have delicate nuances of lemons, apples, cherries, berries, peaches, and other fruits.

TIP

The sparkling wines of the world fall into two broad styles, according to how they're made, and how they taste as a result:

- Wines that express the character of their grapes; these wines tend to be fruity and straightforward, without layers of complexity.
- Wines that express complexity and flavours (yeasty, biscuity, caramel-like, honeyed) that derive from winemaking and aging, rather than expressing overt fruitiness.

How sweet is it?

Nearly all sparkling wines aren't technically dry because they contain measurable but small amounts of sugar, usually as the result of sweetening added at the last stage of production. But all sparkling wines don't necessarily taste sweet. The perception of sweetness depends on two factors: the actual amount of sweetness in the wine (which varies according to the wine's style) and the wine's balance between acidity and sweetness.

Champagne itself is made in a range of sweetness levels, the most common of which is a dry style called *brut*. Sparkling wines made by the *traditional method* used in Champagne are made in the same range of styles as Champagne.

TIP

Inexpensive sparkling wines tend to be medium sweet in order to appeal to a mass market that enjoys sweetness. Wines labeled with the Italian word *spumante* tend to be overtly sweet.

How good is it?

When you taste a sparkling wine, the most important consideration is whether you like it. To evaluate a sparkling wine the way professionals do, however, you have to apply a few criteria that don't apply to still wines (or are less critical in still wines than in sparkling wines):

- **The appearance of the bubbles:** In the best sparkling wines, the bubbles are tiny and float upward in a continuous stream from the bottom of your glass. If the bubbles are large and random, you have a clue that the wine is a lesser quality sparkler. If you don't see many bubbles at all, you could have a bad bottle, a poor or smudged glass, or a wine that may be too old.
- **The feel of the bubbles in your mouth:** The finer the wine, the less aggressive the bubbles feel in your mouth.
- **The balance between sweetness and acidity:** Even if a bubbly wine is too sweet or too dry for your taste, to evaluate its quality, consider its sweetness/acid ratio and decide whether these two elements seem reasonably balanced.
- **The texture:** Traditional-method sparkling wines should be somewhat creamy in texture as a result of their extended lees aging.
- **The finish:** Any impression of bitterness on the finish of a sparkling wine is a sign of low quality.

All That Glitters Is Not Champagne

Champagne, the sparkling wine of Champagne, France, is the gold standard of sparkling wines for a number of reasons:

- Champagne is the most famous sparkling wine in the world; the name has immediate recognition with everyone, not just wine drinkers.
- A particular technique for making sparkling wine was perfected in the Champagne region.
- Champagne is not only the finest sparkling wine in the world, but also among the finest wines in the world of any type.

WARNING!

Within the European Union, only the wines of the Champagne region in France can use the name Champagne. Elsewhere, because of Champagne's fame, the name *champagne* appears on labels of all sorts of sparkling wines that don't come from the Champagne region and that don't taste like Champagne. Many wine drinkers also use the word "champagne" indiscriminately to refer to all wines that have bubbles.

REMEMBER

The word Champagne in this bookazine is referring to true Champagne, from the region of the same name; the generic term *sparkling wine* refers to bubbly wines collectively, and sparkling wines other than Champagne.

Champagne and Its Magic Wines

Champagne, the real thing, comes only from the region of Champagne in northeast France. Champagne is the most northerly vineyard area in France. Most of the important Champagne *houses* (as Champagne producers are called) are located in the cathedral city of Rheims and in the town of Epernay, south of Rheims. Around Rheims and Epernay are the main vineyard areas, where three permitted grape varieties for Champagne flourish. These areas are

- The Montagne de Reims (south of Rheims), where the best Pinot Noir grows
- The Côte des Blancs (south of Epernay), home of the best Chardonnay
- The Valleé de la Marne (west of Epernay), most favourable to Pinot Meunier (a black grape) although all three grape varieties grow there

Most Champagne is made from all three grape varieties — two black and one white. Pinot Noir contributes body, structure, and longevity to the blend; Pinot Meunier provides precocity, floral aromas, and fruitiness; and Chardonnay offers delicacy, freshness, and elegance.

Nonvintage Champagne

Nonvintage (NV) Champagne – any Champagne without a vintage year on the label – accounts for 85 percent of all Champagne. Its typical blend is two-thirds black grapes (Pinot Noir and Pinot Meunier) and one-third white (Chardonnay). Wine from three or more harvests usually goes into the blend.

Each Champagne house blends to suit its own house style for its nonvintage Champagne. (For example, one house may seek elegance and finesse in its wine, another may opt for fruitiness, and a third may value body, power, and longevity.) Maintaining a consistent house style is vital because wine drinkers get accustomed to their favorite Champagne's taste and expect to find it year after year.

Most nonvintage Champagnes sell for £15 to £30 a bottle. Often, a large retailer buys huge quantities of a few major brands, obtaining a good discount that he passes on to his customers. Seeking out stores that do a large-volume business in Champagne is worth your while.

Vintage Champagne

Historically, only in about five of every ten years has the weather in Champagne been good enough to make a Vintage Champagne – that is, the grapes were ripe enough that some wine could be made entirely from the grapes of that year without being blended with reserve wines from previous years. Since 1995, the climate in Champagne (and throughout Europe) has been much warmer than normal, and Champagne producers have been able to make Vintage Champagne almost every year. (The year 2001 was the one exception).

Vintage Champagne is more intense in flavour than nonvintage Champagne. It is typically fuller-bodied and more complex, and its flavours last longer in your mouth. Being fuller and richer, these Champagnes are best with food. Nonvintage Champagnes – usually lighter, fresher, and less complicated – are suitable as apéritifs, and they are good values. Whether a Vintage Champagne is worth its extra cost or not is a judgment you have to make for yourself.

TIP

The Champagne region has had a string of really fine vintages since 1995, especially the 1996 vintage. The three years that followed—1997, 1998, and 1999—all have been good. Both 2000 and 2003 were no more than average (too hot, especially 2003), but 2002 and 2004 are fine vintages (with 2002 the best since 1996), and 2005 is variable. Champagne lovers should seek out 1996 Vintage Champagnes; 1996 is exceptional, one of the best, long-lived vintages ever!

Blanc de blancs and blanc de noirs

A small number of Champagnes derive only from Chardonnay; that type of Champagne is called *blanc de blancs* – literally, "white (wine) from white (grapes)." A blanc de blancs can be a Vintage Champagne or a nonvintage. It usually costs a few dollars more than other Champagnes in its category. Because they're generally lighter and more delicate than other Champagnes, blanc de blancs make ideal apéritifs. Not every Champagne house makes a blanc de blancs. Four of the best all-Vintage Champagnes, are Taittinger Comte de Champagne, Billecart-Salmon Blanc de Blancs, Deutz Blanc de Blancs, and Pol Roger Blanc de Chardonnay.

Rosé Champagne

Rosé Champagnes – pink Champagnes – can also be vintage or nonvintage. Usually, Pinot Noir and Chardonnay are the only grapes used, in proportions that vary from one house to the next.

Colours vary quite a lot, from pale onion-skin to salmon to rosy pink. (The lighter-coloured ones are usually drier.)

Rosés are fuller and rounder than other Champagnes and are best enjoyed with dinner.

TIP

Like blanc de blancs Champagnes, rosés usually cost a few pounds more than regular Champagnes, and not every Champagne house makes one. Some of the best rosés are those of Roederer, Billecart-Salmon, Gosset, and Moët & Chandon (especially its Dom Pérignon Rosé).

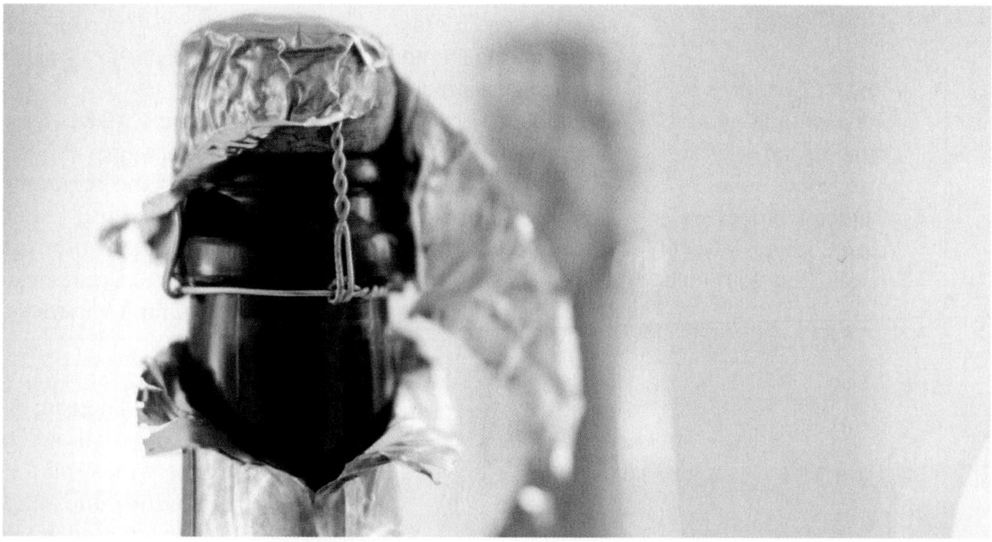

Sweetness categories

Champagnes always carry an indication of their sweetness on the label, but the words used to indicate sweetness are cryptic: Extra dry isn't really dry, for example.

In ascending order of sweetness, Champagnes are labeled as

- Extra brut, brut nature, or brut sauvage: Totally dry
- Brut: Dry
- Extra dry: Medium dry
- Sec: Slightly sweet
- Demi-sec: Fairly sweet
- Doux: Sweet

Recommended Champagne producers

The Champagne business is dominated by about 25 or 30 large houses, most of whom purchase from independent growers the majority of grapes they need to make their Champagne. Of the major houses, only Roederer and Bollinger own a substantial portion of the vineyards from which it gets its grapes – a definite economic and quality-control advantage for them.

TIP

Moët & Chandon is by far the largest Champagne house. In terms of worldwide sales, other large brands are Veuve Clicquot, Mumm, Vranken, Laurent-Perrier, Pommery, Nicolas Feuillate, and Lanson. The following lists name some favorite producers, grouped according to the style of their Champagne: light-bodied, medium-bodied, or full-bodied.

Light, elegant styles

Laurent-Perrier	G.H. Mumm
Taittinger	Bruno Paillard
Ruinart	Perrier-Jouët
Jacquesson	J. Lassalle*
Pommery	Billecart-Salmon
Piper-Heidsieck	

Medium-bodied styles

Charles Heidsieck	Deutz
Pol Roger	Cattier*
Moët & Chandon	Philipponnat

Full-bodied styles

Krug	Alfred Gratien*
Louis Roederer	Delamotte
Bollinger	Salon*
Gosset	Paul Bara*
Veuve Clicquot	

** Small producer; may be difficult to find.*

Bubbling Beauties: Other Sparkling Wines

Wineries all over the world have emulated Champagne by adopting the techniques used in the Champagne region. Their wines differ from Champagne, however, because their grapes grow in terroirs different from that of the Champagne region and because, in some cases, their grapes are different varieties.

Still other sparkling wines are made by using the tank fermentation rather than the bottle fermentation method specifically to attain a certain style, or to reduce production costs.

French sparkling wine

France makes many other sparkling wines besides Champagne, especially in the Loire Valley, around Saumur, and in the regions of Alsace and Burgundy. Sparkling wine made by the traditional method (second fermentation in the bottle) often carries the name *Crémant,* as in Crémant d'Alsace, Crémant de Loire, Crémant de Bourgogne, and so on. Grape varieties are those typical of each region.

Some of the leading brands of French sparkling wines are Langlois-Château, Bouvet Ladubay, Gratien & Meyer (all from the Loire Valley), Brut d'Argent, Kriter, and Saint Hilaire. These wines sell for £6 to £9 and are decent. They're perfect for parties and other large gatherings, when you may want to serve a French bubbly without paying a Champagne price.

American sparkling wine

Almost as many states make sparkling wine as make still wine, but California and New York are the most famous for it. Two fine producers of New York State sparkling wines in the traditional method are Chateau Frank and Lamoreaux Landing, both under £12.

Italian spumante: Sweet or dry

Spumante is simply the Italian word for "sparkling." It often appears on bottles of American wines that are sweet, fruity spin-offs of Italy's classic Asti Spumante. Actually, Italy makes many fine, dry spumante wines and a popular, slightly sparkling wine called Prosecco, as well as sweet spumante"

- **Asti** is a delicious, fairly sweet, exuberantly fruity sparkling wine made in the Piedmont region from Moscato grapes, via the tank method. It's one bubbly that you can drink with dessert.

TIP

Because freshness is essential in Asti, buy a good brand that sells well. (Asti isn't vintage-dated, and so there's no other way to determine how old the wine is.) Try Fontanafredda (about £8 to £9), Martini & Rossi (£7), and Cinzano (about £6).

- Using the traditional method, Italy produces a good deal of dry spumante wine in the Oltrepò-Pavese and Franciacorta wine zones of Lombardy, and in Trentino. Italy's dry sparkling wines are very dry with little or no sweetening dosage. They come in all price ranges.

- A quintessential Italian sparkling wine, **Prosecco** comes from Prosecco grapes grown near Venice and Treviso. It's a straightforward, pleasant apéritif, low in alcohol (about 11 to 12 percent), and it comes in dry, off-dry, and sweet styles. Prosecco is mainly a frizzante wine, but it also comes as a spumante (fully sparkling), or even as a nonsparkling wine (it's better with bubbles).

Prosecco is the perfect wine to have with Italian antipasto, such as pickled vegetables, calamari, anchovies, or spicy salami. Its fresh, fruity flavours cleanse your mouth and get your appetite going for dinner. And Prosecco is eminently affordable: It retails for £7 to £11 a bottle.

Spanish sparkling wines (Cava)

Almost all of Spain's sparkling wine, Cava, which sells mainly for £5 to £7 a bottle, comes from the Penedés region, near Barcelona. Cava is made in the traditional method, fermented in the bottle. But most Cavas use local Spanish grapes. As a result, they taste distinctly different (a nicely earthy, mushroomy flavour) from California bubblies and from Champagne. Some of the more expensive blends do contain Chardonnay.

TIP

Australia, New Zealand, and South Africa now make some very fine sparkling wines in the traditional method. Australia boasts a really good £6 sparkler called Seaview Brut; for something completely unusual, try Seaview's deep red sparkling Shiraz, about £7. Among New Zealand bubblies, one of the finest is that of Highfield Estate in Marlborough.

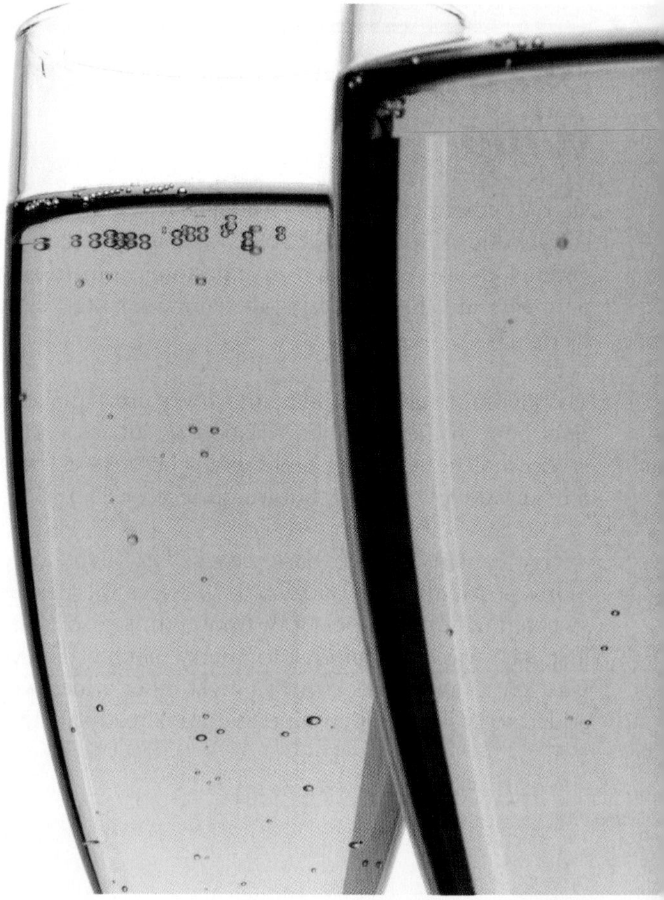

Wine Roads Less Traveled: Fortified and Dessert Wines

IN THIS ARTICLE

- *Introducing Sherry*
- *Meeting Marsala, Vin Santo, and friends*
- *Honoring Port and Madeira*
- *Producing Sauternes*

The wines lumped together as *fortified wines* and *dessert wines* aren't mainstream beverages that you want to drink every day. Some of them are much higher in alcohol than regular wines, and some of them are extremely sweet (and rare and expensive!). They're the wine equivalent of really good chocolate – delicious enough that you can get carried away if you let yourself indulge daily. So you treat them as treats, a glass before or after dinner, a bottle when company comes, a splurge to celebrate the start of your diet – tomorrow.

Sherry: A Misunderstood Wine

Sherry comes from the Andalucía region of sun-baked, southwestern Spain. This area's soil is *albariza,* the region's famous chalky earth, rich in limestone from fossilized shells. Summers are hot and dry, but balmy sea breezes temper the heat.

The Palomino grape – the main variety used in Sherry – thrives only here in the hot Sherry region on albariza soil. Palomino is a complete failure for table wines because it is so neutral in flavour and low in acid, but it's perfect for Sherry production.

Sherry consists of two basic types: *fino* (light, very dry) and *oloroso* (rich and full, but also dry). Sweet Sherries are made by sweetening either type. Both fino and oloroso Sherries age in a special way that's unique to Sherry making. The young wine isn't left to age on its own (as most other wines would) but is added to casks of older wine that are already aging.

WARNING!

Authentic Sherry is made only in the Jerez region of Spain and carries the official name, *Jerez-Xérès-Sherry* (the Spanish, French, and English names for the town) on the front or back label.

Among dry Sherries, these are the main styles:

- **Fino:** Pale, straw-coloured Sherry, light in body, dry, and delicate. Fino Sherries are always matured under flor (a film of yeast on the surface of the wine), either in Jerez or Puerto de Santa María. They have 15 to 17 per cent alcohol. After they lose their protective flor (by bottling), finos become very vulnerable to oxidation spoilage, and you must therefore store them in a cool place, drink them young, and refrigerate them after opening. They're best when chilled.

- **Manzanilla:** Pale, straw-coloured, delicate, light, tangy, and very dry fino-style Sherry made only in Sanlucar de Barrameda. (Although various styles of manzanilla are produced, *manzanilla fina,* the fino style, is by far the most common.) The temperate sea climate causes the flor to grow thicker in this town, and manzanilla is thus the driest and most pungent of all the Sherries. Handle it similarly to a fino Sherry.

- **Manzanilla pasada:** A manzanilla that has been aged in cask about seven years and has lost its flor. It's more amber in colour than a manzanilla fina and fuller-bodied. It's close to a dry amontillado (see the next item) in style, but still crisp and pungent. Serve cool.

- **Amontillado:** An aged fino that has lost its flor in the process of cask aging. It's deeper amber in colour and richer and nuttier than the previous styles. *Amontillado (ah moan tee YAH doh)* is dry but retains some of the pungent tang from its lost flor. True amontillado is fairly rare; most of the best examples are in the £15 to £25 price range. Cheaper Sherries labeled "amontillado" are common, so be suspicious if it costs less than £9 a bottle. Serve amontillado slightly cool and, for best flavour, finish the bottle within a week.

- **Oloroso:** Dark gold to deep brown in colour (depending on its age), full-bodied with rich, raisiny aroma and flavour, but dry. Olorosos lack the delicacy and pungency of fino (flor) Sherries. Serve them at room temperature.

- **Palo cortado:** The rarest of all Sherries. It starts out as a fino, with a flor, and develops as an amontillado, losing its flor. But then, for some unknown reason, it begins to resemble the richer, more fragrant oloroso style, all the while retaining the elegance of an amontillado. In colour and alcohol content, palo cortado *(PAH loe cor TAH doh)* is similar to an oloroso, but its aroma is quite like an amontillado. Like amontillado Sherry, beware of cheap imitations. Serve at room temperature. It keeps as well as olorosos.

Serving and storing Sherry

The light, dry Sherries – fino and manzanilla – must be fresh. Buy them from stores with rapid turnover; a fino or manzanilla that has been languishing on the shelf for several months will not give you the authentic experience of these wines.

Although fino or manzanilla can be an excellent apéritif, be careful when ordering a glass in a restaurant or bar. Never accept a glass from an already-open bottle unless the bottle has been refrigerated. Even then, ask how long it has been open – more than two days is too much. After you open a bottle at home, refrigerate it and finish it within a couple of days.

Buy half-bottles of fino and manzanilla so that you don't have leftover wine that oxidizes. These, and all Sherries, can be stored upright. Try not to hold bottles of fino or manzanilla more than three months, however. The higher alcohol and the oxidative aging of other types of Sherry (amontillado, oloroso, palo cortado, all the sweet Sherries) permit you to hold them for several years.

Manzanilla and fino Sherry are ideal with almonds, olives, shrimp or prawns, all kinds of seafood, and those wonderful tapas in Spanish bars and restaurants. Amontillado Sherries can accompany tapas before dinner but are also fine at the table with light soups, cheese, ham, or salami (especially the Spanish type, *chorizo*). Dry olorosos and palo cortados are best with nuts, olives, and hard cheeses (such as the excellent Spanish sheep-milk cheese, Manchego). All the sweet Sherries can be served with desserts after dinner or enjoyed on their own.

Port: The Glory of Portugal

Port is the world's greatest fortified red wine. Port wine is fermented and fortified in the Douro Valley, and then most of it travels downriver to the coast.

TIP

Don't let all the complicated styles of Port deter you from picking up a bottle and trying it. If you've never had Port before, you're bound to love it – almost no matter which style you try. (Later, you can fine-tune your preference for one style or another.) Port is, simply, delicious!

Although all Port is sweet, and most of it is red, a zillion styles exist. The styles vary according to the quality of the base wine (ranging from ordinary to exceptional), how long the wine is aged in wood before bottling (ranging from 2 to 40-plus years), and whether the wine is from a single year or blended from wines of several years.

Following is a brief description of the main styles, from simplest to most complex:

- **White Port:** Made from white grapes, this gold-coloured wine can be off-dry or sweet. You may wonder why white Port exists – Sherries and Sercial Madeiras are better as apéritifs and red Ports are far superior as sweet wines. But try a white Port with tonic and ice one day, and you'll have your answer! Served this way, white Port can be a bracing warm-weather apéritif.

- **Ruby Port:** This young, nonvintage style is aged in wood for about three years before release. Fruity, simple, and inexpensive (around £7 for major brands), it's the best-selling type of Port. If labeled *Reserve* or *Special Reserve,* the wine has usually aged about six years and costs a few dollars more. Ruby Port is a good introduction to the Port world.

- **Vintage Character Port:** Despite its name, this wine isn't single-vintage Port – it just tries to taste like one. Vintage Character Port is actually premium ruby blended from higher-quality wines of several vintages and matured in wood for about five years. Full-bodied, rich, and ready-to-drink when released, these wines are a good value at about £10 to £11. But the labels don't always say *Vintage Character;* instead, they often bear proprietary names such as Founder's Reserve (from Sandeman); Bin 27 (Fonseca); Six Grapes (Graham); First Estate (Taylor Fladgate); Warrior (Warre); and Distinction (Croft). As if *Vintage Character* wouldn't have been confusing enough!

- **Tawny Port:** Tawny is the most versatile Port style. The best tawnies are good-quality wines that fade to a pale garnet or brownish red colour during long wood aging. Their labels carry an indication of their average age (the average age of the wines from which they were blended) – 10, 20, 30, or over-40 years. Ten-year-old tawnies cost about £18, 20-year-olds sell for £28 to £31, and 30- and over-40-year-old tawnies cost a lot more (£56 to well over £60). Ten- and 20-year-old tawnies are the best buys; the older ones aren't always worth the extra dosh. Tawny Ports have more finesse than other styles and are appropriate both as apéritifs and after dinner. Inexpensive tawnies that sell for about the same price as ruby Port are usually weak in flavour and not worth buying.

TIP

You can enjoy a serious tawny Port in warm weather (even with a few ice cubes!) when a Vintage Port would be too heavy and tannic.

- **Colheita Port:** Often confused with Vintage Port because it's vintage-dated, colheita is actually a tawny from a single vintage. In other words, it has aged (and softened and tawnied) in wood for many years. Unlike an aged tawny, though, it's the wine of a single year. Niepoort is one of the few Port houses that specializes in colheita Port. It can be very good but older vintages are quite expensive (about £60 plus). Smith Woodhouse and Delaforce offer colheita Portos for £30 or less.

- **Late Bottled Vintage Port (LBV):** This type *is* from a specific vintage, but usually not from a very top year. The wine ages four to six years in wood before bottling and is then ready to drink, unlike Vintage Port. Quite full-bodied, but not as hefty as Vintage Port, it sells for about £11 to £14.

- **Vintage Port:** The pinnacle of Port production, Vintage Port is the wine of a single year blended from several of a house's best vineyards. It's bottled at about two years of age, before the wine has much chance to shed its tough tannins. It therefore requires an enormous amount of bottle aging to accomplish the development that didn't occur in wood. Vintage Port is usually not mature (ready to drink) until about 20 years after the vintage.

  Most good Vintage Ports sell for £45 to £60 when they're first released (years away from drinkability). Mature Vintage Ports can cost well over £60. Producing a Vintage Port amounts to a *declaration of that vintage* (a term you hear in Port circles) on the part of an individual Port house.

Storing and serving Port

Treat Vintage Ports like all other fine red wines: Store the bottles on their sides in a cool place. You can store other Ports either on their sides (if they have a cork rather than a plastic-topped cork stopper) or upright. All Ports, except white, ruby, and older Vintage Ports, keep well for a week or so after opening, with aged-stated tawny capable of keeping for a few weeks.

Serve Port at cool room temperature, 64°F (18°C), although tawny Port can be an invigorating pick-me-up when served chilled during warm weather. The classic complements to Port are walnuts and strong cheeses, such as Stilton, Gorgonzola, Roquefort, Cheddar, and aged Gouda.

WARNING!

Because it's very rich and very tannic, this wine throws a heavy sediment and *must* be decanted, preferably several hours before drinking (it needs the aeration). Vintage Port can live 70 or more years in top vintages.

- **Single Quinta Vintage Port:** These are Vintage Ports from a single estate *(quinta)* that is usually a producer's best property (such as Taylor's Vargellas and Graham's Malvedos). They're made in good years, but not in the best vintages, because then their grapes are needed for the Vintage Port blend. They have the advantage of being readier to drink than declared Vintage Ports – at less than half their price — and of usually being released when they're mature. You should decant and aerate them before serving, however.

Marsala, Vin Santo, and the Gang

Italy has a number of interesting dessert and fortified wines, of which Marsala (named after a town in western Sicily) is the most famous. Marsala is a fortified wine made from local grape varieties. It comes in numerous styles, all of which are fortified after fermentation, like Sherry. You can find dry, semi-dry, or sweet versions and amber, gold, or red versions, but the best Marsalas have the word *Superiore* or – even better – *Vergine* or *Vergine Soleras* on the label. Marsala Vergine is unsweetened and uncoloured, and is aged longer than other styles.

The region of Tuscany is rightfully proud of its Vin Santo *(vin SAHN toh)*, a golden amber wine made from dried grapes and barrel-aged for several years. Vin Santo can be dry, medium-dry, or sweet. The dry style is nice as an apéritif, and the medium-dry version works well as an accompaniment to the wonderful Italian almond cookies called *biscotti*.

TIP

Many Tuscan producers make a Vin Santo; four outstanding examples of Vin Santo (conveniently available in half-bottles as well as full bottles) are from Avignonesi (very expensive!), Badia a Coltibuono, Castello di Cacchiano, and San Giusto a Rentennano.

Long Live Madeira

The legendary wine called Madeira comes from the island of the same name, which sits in the Atlantic Ocean nearer to Africa than Europe. Madeira is a subtropical island whose precarious hillside vineyards rise straight up from the ocean. The island is a province of Portugal, but the British have always run its wine trade.

The very best Madeira wines are still those from the old days, vintage-dated wines from 1920 back to 1795. Surprisingly, you can still find a few Madeiras from the nineteenth century. The prices aren't outrageous, either (£200 to £300 a bottle), considering what other wines that old, such as Bordeaux, cost.

The best Madeira comes in four styles, two fairly dry and two sweet. The sweeter Madeiras generally have their fermentation halted somewhat early by the addition of alcohol. Drier Madeiras have alcohol added after fermentation.

You never have to worry about Madeira getting too old. It's indestructible. The enemies of wine – heat and oxygen – have already had their way with Madeira during the winemaking and maturing process. Nothing you do after it's opened can make it blink.

Sauternes and the Nobly Rotten Wines

Warm, misty autumns encourage the growth of a fungus called *botrytis cinerea* in vineyards. Nicknamed *noble rot,* botrytis concentrates the liquid and sugar in the grapes, giving the winemaker amazingly rich juice to ferment. The best wines from botrytis-infected grapes are among the greatest dessert wines in the world, with intensely concentrated flavours and plenty of acidity to prevent the wine from tasting excessively sweet.

The greatest nobly rotten wines are made in the Sauternes district of Graves (Bordeaux) and in Germany, but they are also produced in Austria and California, among other places.

Sauternes is a very labour-intensive wine. Consequently, good Sauternes is expensive. Prices range from £28 to £31 a bottle up to £188 (depending on the vintage) for Château d'Yquem *(d'ee kem).*

TIP

Sauternes is widely available in half-bottles, reducing the cost somewhat. A 375 ml bottle is a perfect size for after dinner, and you can buy a decent half-bottle of Sauternes or Barsac, a dessert wine similar to Sauternes, like Château Doisy-Védrines *(dwahs ee veh dreen)* or Château Doisy-Daëne *(dwahs ee dah en)* for £15 to £18.

Sauternes is best when served cold, but not ice cold, at about 52° to 53°F (11°C). Mature Sauternes can be served a bit warmer. Because the wine is so rich, Sauternes is an ideal companion for foie gras although, ordinarily, the wine is far more satisfying after dinner than as an apéritif. As for desserts, Sauternes is excellent with ripe fruits, lemon-flavored cakes, or pound cake.

TIP

Sauternes has such balance of natural sweetness and acidity that it can age well for an extraordinarily long time. Unfortunately, because Sauternes is so delicious, people often drink it young, when it's very rich and sweet. But Sauternes is really at its best when it loses its baby fat and matures.

Vintage Wine Chart: 1995–2004

Key:

- 100 = Outstanding
- 95 = Excellent
- 90 = Very Good
- 85 = Good
- 80 = Fairly Good
- 75 = Average
- 70 = Below Average
- 65 = Poor
- 50 – 60 = Very Poor
- a = Too young to drink
- b = Can be consumed now, but will improve with time
- c = Ready to drink
- d = May be too old
- NV = Non-vintage year

WINE REGION	1995	1996	1997	1998	1999	2000	2001	2002	2003	2004
Bordeaux										
Médoc, Graves	90b	90a	85b	85a	85a	95a	85a	85b	85a	85a
Pomerol, St-Emil	90b	85a	85b	95a	85a	95a	85a	85b	80a	85a
Sauternes/Barsac	85b	85b	85b	85a	85a	85a	95a	90a	90a	80a
Burgundy										
Côte de Nuits	90b	90a	90c	80c	90b	85c	80b	90a	85a	85a
Côte Beaune	85b	90b	85c	80c	90b	75c	75c	90a	80b	80b
Burgundy White	90b	90b	85c	80c	85c	85c	85c	90b	80b	85b
Rhône Valley										
Northern Rhône	90b	85c	90c	90a	95a	85b	90a	75c	95a	80c
Southern Rhône	90b	80d	80d	95b	90b	95b	95a	60d	90b	85c
Other Wine Regions										
Alsace	85c	85c	85c	90c	85c	90c	90c	85c	75c	90b
Champagne	85b	100a	85b	85c	85c	80c	NV	90a	?	?
Germany	85c	90b	85c	90b	85c	70d	90b	90b	85b	90a
Rioja	90c	85c	85c	80d	85c	85c	95b	75d	85c	90b
Vintage Port	NV	NV	85a	NV	NV	90a	NV	NV	90a?	90a?
Italy										
Piedmont	85b	95a	85c	90a	95a	90b	95a	75c	80b	85a?
Tuscany	85b	75c	90c	85c	95c	85c	95b	75c	80b	85a?
California North Coast										
Cab.Sauvignon	90b	90b	90c	85c	85b	75c	95a	90b	90b	85b

Wine Region	Recent Past Great Vintages (prior to 1985)
Bordeaux	
Médoc, Graves	1959, 1961, 1970, 1982
Pomerol, St-Emil	1961, 1964, 1970, 1975, 1982
Burgundy	
Côte de Nuits-Red	1959, 1964, 1969, 1978
Côte Beune-Red	1959, 1969
Burgundy, White	1962, 1966, 1969, 1973, 1978
Rhone Valley	
Northern Rhône	1959, 1961, 1966, 1969, 1970, 1972 (Hermitage), 1978
Southern Rhône	1961, 1967, 1978

Wine Region	Recent Past Great Vintages (prior to 1985)
Other Wine Regions	
Alsace	1959, 1961, 1967, 1976, 1983
Champagne	1961, 1964, 1969, 1971, 1975, 1979, 1982
Sauternes	1959, 1962, 1967, 1975, 1983
Rioja (Spain)	1964, 1970, 1981, 1982
Vintage Port	1963, 1966, 1970, 1977, 1983
Italy	
Piedmont	1958, 1964, 1971, 1978, 1982
Tuscany	1967, 1970 (Brunello di Montalicino), 1971
California North Coast	
Cabernet Sauvignon	1951, 1958, 1968, 1970, 1974, 1978